1983

Time Series Models

A C HARVEY

The London School of Economics

A HALSTED PRESS BOOK

John Wiley & Sons
New York

First published 1981 by
PHILIP ALLAN PUBLISHERS LIMITED
MARKET PLACE
DEDDINGTON
OXFORD OX5 4SE

Published in the USA by
HALSTED PRESS, A DIVISION OF
JOHN WILEY & SONS INC.,
NEW YORK

ISBN 0-470-27259-7

British Library Cataloguing in Publication Data
Harvey, A. C.
    Time series models.
    1. Econometrics   2. Time series analysis
    I. Title
    330'.01'51955   HB 139

To Catherine and Samuel

Contents

Preface

This book is concerned with the analysis and modelling of time series. It is designed primarily for use in courses given to final year undergraduates and postgraduates in statistics and econometrics. Although the emphasis is on economic time series, the material presented is also relevant in engineering and geography, and in other disciplines where time series observations are important.

'Time Series Models' can be regarded as a companion volume to my earlier book, 'The Econometric Analysis of Time Series'. The two books are essentially self-contained, although there is some cross-referencing. Here, reference to the earlier book will be indicated by the rather tasteless abbreviation 'EATS'.[1]

As in EATS, the main concern in this book is to concentrate on models and techniques which are of practical value. There is more stress on creating an understanding of what the various models are capable of, and the ways in which they can be applied, than in proving theorems with the maximum of mathematical rigour. It is assumed that the reader is familiar with calculus and matrix algebra, although a good deal of the book can be read without any knowledge of matrices whatsoever. A basic knowledge of statistical inference is also assumed. Thus the student who has studied, say, Silvey (1970) or Chapters 3, 4 and 5 of EATS will be fully equipped to read the sections on estimation and inference. A summary of the key results is provided in the introductory chapter.

A short course in time series model building and forecasting could be based on Chapters 1, 2, 5 and 6, with the starred sections omitted.

[1] For a summary of the contents of EATS, see page x.

A more comprehensive course on time series analysis would almost
certainly include the material in spectral analysis in Chapter 3,
although individual teachers might want to be selective in their
choice of topics in Chapters 4 and 7. The reason for including
Chapter 4 is that there is a growing interest in state space models and
their relationship to more conventional time series techniques.
Unfortunately, most of the accounts of state space models currently
available are in the engineering literature. The way the material is
presented in Chapter 4 will, I hope, make it more accessible to those
with a background in statistics or econometrics.

I have defined time series models as models which are constructed
without drawing on any theories concerning possible behavioural
relationships between variables. In univariate analysis the movements
of a variable are explained solely in terms of its own past and its
position with respect to time. In multivariate analysis the same thing
is done, but with respect to a set of variables. The actual decision to
model several variables jointly does, in fact, imply the use of some
prior knowledge, but the important point is that no *a priori*
restrictions are placed on the form of the model.

The decision to include a chapter on regression models in a sense
contradicts the previous paragraph. However, only static models are
considered and the form in which the explanatory variables enter the
model is taken for granted. The reason for including the chapter is to
focus attention on certain aspects of regression, peculiar to time
series, which are not normally included in texts on the general linear
model.

Equations are numbered according to the section. The chapter
number is omitted except when referring to an equation in another
chapter. Examples are numbered within each section and are
referenced in the same way as equations. Tables and figures are
numbered consecutively throughout each chapter and are independent of the section in which they appear. The 'Notes' at the end of
each chapter are primarily for references not given in the text, and
for further reading.

As in EATS, certain sections are starred (*). These sections
contain material which is more difficult or more esoteric, or both.
They can be omitted without any loss of continuity, although for a
graduate course most of them would be included.

Bold face type-setting is not used in the text. As a general rule
lower case letters denote scalars or vectors, while, with a few obvious
exceptions, upper case letters indicate matrices. This convention is
violated significantly only in Chapter 3 where capital letters are used
to denote scalars in a number of cases; matrices are used only in the
first part of Section 7.

The term 'log' denotes a natural logarithm unless explicitly stated otherwise.

Parts of the book have been used as the basis for lectures at the LSE, and I'm grateful to all the students whose questions forced me to think more clearly about the exposition of certain topics. I'm also grateful to all the colleagues and friends who were kind enough to comment on various drafts of the book. Special thanks must go to Dick Baillie, Tom Cooley, James Davidson, Rob Engle, Katarina Juselius, Colin McKenzie and Bianca De Stavola. Of course, I am solely responsible for any errors which may remain. Finally, I'd like to thank Jill Duggan, Hazel Rice, Sue Kirkbride and Maggie Robertson for typing a difficult manuscript so efficiently.

London
February 1981

Note

The companion volume by A. C. Harvey, *The Econometric Analysis of Time Series* (Philip Allan 1981) has the following contents:

1
Introduction

1. Analysing and Modelling Time Series

A time series typically consists of a set of observations on a variable, y, taken at equally spaced intervals over time. Economic variables are generally classified either as stocks or flows, the money supply being an example of a stock, and investment and gross national product being flows. The distinction between stocks and flows is important in dealing with aggregation or missing observations, but for most purposes it is irrelevant, and a series of T observations will be denoted by $y_1, \ldots, y_T$, irrespective of whether they refer to a stock or a flow variable.

There are two aspects to the study of time series — analysis and modelling. The aim of analysis is to summarise the properties of a series and to characterise its salient features. This may be done either in the time domain or in the frequency domain. In the time domain attention is focused on the relationship between observations at different points in time, while in the frequency domain it is cyclical movements which are studied. The two forms of analysis are complementary rather than competitive. The same information is processed in different ways, thereby giving different insights into the nature of the time series.

The main reason for modelling a time series is to enable forecasts of future values to be made. The distinguishing feature of a time series model, as opposed, say, to an econometric model, is that no attempt is made to relate y_t to other variables. The movements in y_t are 'explained' solely in terms of its own past, or by its position in relation to time. Forecasts are then made by extrapolation.

1

The Nature of Time Series Models

Figure 1.1 shows a series of observations fluctuating around a fixed level, μ. If the observations were independent of each other, the best forecast of the next observation in the series, y_{T+1}, would simply be μ or, if this is unknown, a suitable estimate of μ such as the sample mean. However, the observations are clearly not independent. Each one tends to have a value which is closer to that of the observations immediately adjacent, than to those which are further away. This type of structure is known as *serial correlation*. It is typical of time series observations and by taking account of the pattern of serial correlation, better forecasts of future observations can be obtained. Thus, given the relationship between successive observations in figure 1.1, it would seem that a forecast lying somewhere between y_T and μ would be more appropriate than a forecast which is simply equal to μ.

The statistical approach to forecasting is based on the construction of a model. The model defines a mechanism which is regarded as being capable of having produced the observations in question. Such a model is almost invariably *stochastic*. If it were used to generate several sets of observations over the same time period, each set of observations would be different, but they would all obey the same probabilistic laws.

The first-order autoregressive model

$$y_t - \mu = \phi(y_{t-1} - \mu) + \epsilon_t \tag{1.1}$$

is a simple example of a *stochastic process*. The uncertainty derives

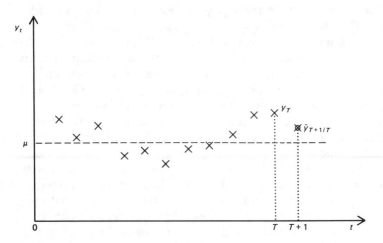

Figure 1.1   *Time Series with Serial Correlation*

from the variable ϵ_t. This is a purely random disturbance term with a mean of zero and a variance of σ^2. It is purely random in the sense that the correlation between any two of its values at different points in time is zero. The remaining features of the model are determined by the parameters μ and ϕ. If $|\phi| < 1$, the observations fluctuate around μ, which is then the mean of the process.

The parameters in (1.1) can be estimated by *ordinary least squares* (OLS) regression. Given these estimates, $\tilde{\mu}$ and $\tilde{\phi}$, the next observation in the series can be forecast by

$$\tilde{y}_{T+1/T} = \tilde{\mu} + \tilde{\phi}(y_T - \tilde{\mu}) \tag{1.2}$$

The closer $\tilde{\phi}$ is to one, the more weight is given to y_T. This is consistent with the intuitive argument given in connection with the observations in figure 1.1.

Constraining ϕ to lie between -1 and 1 in (1.1) means that the process is *stationary*. When a series of observations are generated by a stationary process, they fluctuate around a constant level and there is no tendency for their spread to increase or decrease over time. These are not the only properties of a stationary time series, but they are the most obvious ones and a casual inspection of the series in figure 1.1 indicates that it displays these characteristics.

Further lagged values, y_{t-2}, y_{t-3} and so on, could be added to (1.1), thereby enabling more complicated patterns of dependence to be modelled. Indeed, the properties of almost any stationary time series can be reproduced by introducing a sufficiently high number of lags. The disadvantage of modelling a series in this way is that when a large number of lagged values are needed, a large number of parameters must be estimated. The solution to this problem is to widen the class of models to allow for lagged values of the ϵ_t's. A model which contains lagged values of both the observed variable and the disturbance term is known as an *autoregressive-moving average* (ARMA) process. Such processes play a central role in dynamic modelling because they allow a *parsimonious* representation of a stationary time series. In other words, a model with relatively few parameters can be constructed.

A very simple modification to (1.1) is to introduce a single lagged value of ϵ_t. Thus

$$y_t - \mu = \phi(y_{t-1} - \mu) + \epsilon_t + \theta\epsilon_{t-1}, \tag{1.3}$$

where θ is a moving average parameter. This model is an ARMA process of order $(1, 1)$. It can be denoted simply by writing $y_t \sim \text{ARMA}(1, 1)$.

Although very few actual time series are stationary, the notion of stationarity is fundamental to time series analysis. Stationary ARMA

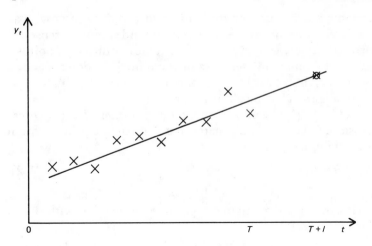

(a) *Forecasting with a global trend model*

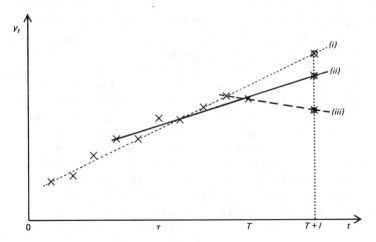

(b) *Forecasting with (i) a global trend; (ii) a local trend; (iii) an extreme local trend*

Figure 1.2   *Local and Global Linear Trends*

processes form the basis upon which more general models can be built. Furthermore, their use is not restricted to pure time series models. The disturbance term in any regression model which uses time series data can be conceived of as a stationary stochastic process. This idea has been widely adopted in modelling behavioural relationships, particularly in engineering and econometrics.

Trends

The majority of time series do not fluctuate around a constant level, but instead show some kind of systematic upward or downward movement. Consider the observations in figure 1.2(a). They can be regarded as being scattered randomly around an upward sloping straight line. Thus a suitable model might be

$$y_t = \alpha + \beta t + \epsilon_t, \qquad t = 1, \ldots, T, \tag{1.4}$$

where α and β are parameters and ϵ_t is a 'white noise' process of the kind defined in (1.1).

Since (1.4) is a classical linear regression model, its parameters can be estimated by OLS. Forecasts of future values of y_t can be made by extrapolating the deterministic part of the fitted equation. Hence

$$\tilde{y}_{T+l/T} = a + b(T + l) = a + bT + bl, \qquad l = 1, 2, \ldots, \tag{1.5}$$

where a and b denote the least squares estimators of α and β. A more general model could be formed by letting the disturbance term follow an ARMA(p, q) process. If such a model were constructed, the forecast in (1.5) would have to be modified to take account of the pattern of serial correlation.

The disadvantage of using (1.4) for forecasting purposes is that the trend is a *global* one. It is assumed to hold at all points in time with the parameters remaining constant throughout. Thus a situation like that shown in figure 1.2(b), where the slope falls after $t = \tau$, cannot be properly handled by (1.4). The question of whether the disturbance term is assumed to be white noise or a more general stationary ARMA process is irrelevant.

A more satisfactory forecasting procedure for the series in figure 1.2(b) would allow the parameters to adapt to the change in the data. If relatively more weight is placed on the most recent observations, forecasts can be based on an estimate of the *local* trend. The lines drawn in figure 1.2(b) contrast the two approaches. Taking this argument to its extreme would mean projecting a line from the last two observations only, i.e.

$$\hat{y}_{T+l/T} = y_T + (y_T - y_{T-1})l, \qquad l = 1, 2, \ldots \tag{1.6}$$

The *Holt–Winters method* is a general procedure for constructing forecasts from a local straight line trend. The forecast function is

$$\hat{y}_{T+l/T} = \hat{\alpha}_T + \hat{\beta}_T l, \qquad l = 1, 2, \ldots, \tag{1.7}$$

where $\hat{\alpha}_T$ and $\hat{\beta}_T$ are calculated from the recursive formulae

$$\hat{\alpha}_t = \lambda_0 y_t + (1 - \lambda_0)(\hat{\alpha}_{t-1} + \hat{\beta}_{t-1}), \tag{1.8a}$$

$$\hat{\beta}_t = \lambda_1(\hat{\alpha}_t - \hat{\alpha}_{t-1}) + (1 - \lambda_1)\hat{\beta}_{t-1}, \qquad t = 3, \ldots, T, \qquad (1.8b)$$

with starting values $\hat{\alpha}_2 = y_2$, $\hat{\beta}_2 = y_2 - y_1$. The *smoothing constants*, λ_0 and λ_1, lie in the range $0 < \lambda_0, \lambda_1 \leqslant 1$. Setting $\lambda_0 = \lambda_1 = 1$ yields forecasts based on (1.6), but as both parameters tend towards zero the forecasts move closer to those given by the global trend model. Note that $\hat{\alpha}_T$ corresponds to $a + bT$ in (1.5).

The values of λ_0 and λ_1 can be fixed by the user on *a priori* grounds. The closer these values are to unity, the more quickly the system adapts to a change in direction, and the more emphasis there is on forecasting accuracy in the very short run.

As it stands, the Holt–Winters system is essentially *ad hoc* and a natural question to ask is whether there is a model which produces forecasts of this form. In the extreme case $\lambda_0 = \lambda_1 = 1$, the answer is straightforward. Let the first differences of the observations be defined by

$$\Delta y_t = y_t - y_{t-1}. \qquad (1.9)$$

Differencing again gives

$$\Delta^2 y_t = y_t - y_{t-1} - (y_{t-1} - y_{t-2}) = y_t - 2y_{t-1} + y_{t-2}, \quad (1.10)$$

and if the assumed model is

$$\Delta^2 y_t = \epsilon_t, \qquad t = 3, \ldots, T, \qquad (1.11)$$

where ϵ_t is white noise, the appropriate forecast of y_{T+1} is

$$\tilde{y}_{T+1/T} = 2y_T - y_{T-1}. \qquad (1.12)$$

This is identical to the forecast produced by (1.6) for $l = 1$. Forecasts for $l > 1$ can be obtained from the difference equation

$$\tilde{y}_{T+l/T} = 2\tilde{y}_{T+l-1/T} - \tilde{y}_{T+l-2/T}, \qquad l = 1, 2, \ldots, \qquad (1.13)$$

with $\tilde{y}_{T/T} = y_T$ and $\tilde{y}_{T-1/T} = y_{T-1}$. This has the solution (1.6) for any value of l.

More generally, a model in which the second differences of the observations are assumed to follow an MA(2) process will give rise to forecasts which can be generated by the Holt–Winters method. If θ_1 and θ_2 are the moving average parameters in

$$\Delta^2 y_t = \epsilon_t + \theta_1 \epsilon_{t-1} + \theta_2 \epsilon_{t-2} \qquad (1.14)$$

then, in terms of (1.8), $\lambda_0 = 1 - \theta_2$ and $\lambda_1 = 1 + \theta_1 + \theta_2$. This suggests the possibility of estimating λ_0 and λ_1 by fitting (1.14) to the observations.

The stochastic process in (1.14) is a special case within a class of models known as *autoregressive-integrated-moving average* (ARIMA) processes. In an ARIMA(p, d, q) process the dth difference of the observations is modelled as an ARMA(p, q) process. Thus (1.14) is ARIMA(0, 2, 2). The fact that such models are capable of picking up local trends is one of the reasons why they are attractive for forecasting.

The approach to time series forecasting based on ARIMA models was developed by Box and Jenkins (1976). The first step is to obtain a series which is approximately stationary. This is normally done by applying transformations, such as logarithms, and by taking first or second differences. The resulting series is then analysed, and on the basis of its properties a suitable ARMA model is chosen. This model is then fitted, and if an examination of the residuals suggests that the model is satisfactory, it is used for forecasting. If, on the other hand, this 'diagnostic checking' indicates serious inadequacies in the specification, an alternative formulation is developed and the whole process is repeated.

Forecasting based on fitting ARIMA processes is open to certain criticisms, one of which is that it requires a good deal of skill to select a suitable model. On the other hand, a procedure such as Holt—Winters can be applied in a semi-automatic fashion. However, even if this point is accepted, the advantages of studying ARIMA processes are undeniable. In the first place, most semi-automatic procedures can be shown to correspond to particular underlying stochastic processes, and a recognition of this point reveals their merits and defects more clearly. In the second place, the mean square error of predictions from an ARIMA model can be evaluated explicitly. This provides a scientific basis upon which to assess forecasting reliability.

Multivariate Time Series

The discussion so far has been confined to the analysis and modelling of a *univariate* time series. When several series are considered together, y_t is no longer regarded as a single observation, but as a vector. The sequence $y_1, \ldots, y_T$, is then a *multivariate* time series.

The decision to model several series jointly implies the availability of *a priori* knowledge which suggests that they are linked together. However, as in univariate time series modelling, no *a priori* knowledge concerning possible behavioural relationships is employed. If y_t is regarded as a stochastic process, each element in the vector of observations is assumed to depend on its own past values and on the past values of all the other variables in the system. If only first order lags are relevant, a generalisation of (1.1) is obtained. This may be written as

$$y_t = \Phi y_{t-1} + \epsilon_t. \tag{1.15}$$

If y_t is an $N \times 1$ vector, the random disturbance term, ϵ_t, is also an $N \times 1$ vector, while Φ is an $N \times N$ matrix of parameters. The disturbances are uncorrelated over time, but may be contemporaneously correlated. Thus $E(\epsilon_t \epsilon_t') = \Omega$, where Ω is an $N \times N$ matrix. As in the univariate case, a model of the form (1.15) is usually only fitted to variables which are stationary. Thus in writing down (1.15) it is to be understood that y_t may denote a set of variables after first or second differences and logarithms have been taken. A non-zero mean may also be incorporated within y_t.

The relationship between multivariate time series models and the systems of dynamic simultaneous equations employed in econometrics is explored in the last chapter of EATS. Although a model of the form (1.15) is often used for forecasting in econometrics, economic theory will typically place *a priori* restrictions on the elements of Φ.

Outline of the Book

Chapter 2 develops the basic concepts needed for the analysis of a stochastic process in the time domain. The rationale behind the ARMA class of models is explained in detail and the time domain properties of various members of the class are examined. In the final sub-section these concepts are extended to multivariate processes.

Chapter 3 is concerned with analysis in the frequency domain. This is sometimes known as *spectral analysis*. The frequency domain properties of various ARMA models are examined and contrasted with the properties of corresponding models in the time domain. It is then shown how spectral analysis can give important insights into the effect of operations which are typically applied to time series. Seasonal adjustment is just one example.

Chapter 4 is an introduction to state space models and the Kalman filter. State space models have been developed primarily in connection with control engineering and it is only recently that their importance has begun to be recognised more widely. As well as providing the technology for the treatment of unobserved components and time-varying parameter models, state space provides the basis for what seems to be a very natural approach to the full maximum likelihood estimation of ARMA models. This topic is pursued in Chapter 5. However, the main emphasis in Chapter 5 is on an approximation to ML estimation which involves the minimisation of what is known as a 'conditional sum of squares' function. Test procedures are also developed in Chapter 5, and in Chapter 6 it is shown how these procedures can be incorporated into a general strategy for model selection. Chapter 6 also deals with the mechanics of making predictions and calculating the associated mean square errors.

The last chapter is concerned with various aspects of time series regression. This might seem out of place in view of the earlier statement that the essential feature of time series models is that they do not involve behavioural relationships. However, the systematic part of the regression model is only of subsidiary interest in this treatment. Its specification is taken for granted and attention is focused on the disturbance term and on methods for modelling parameters which may change over time.

2. Estimation

Most of the methods used for estimating the parameters in time series models are based on the principle of maximum likelihood. The advantage of adopting a maximum likelihood procedure is that the resulting estimators are statistically efficient in large samples. On the other hand, some kind of iterative procedure will usually be necessary to compute them. An introduction to various methods of numerical optimisation will be found in EATS (Chapter 4), although one procedure, Gauss–Newton, is described in some detail below. The emphasis on Gauss–Newton is partly pedagogic, in that it gives some theoretical insight into estimation and testing in a non-linear framework.

The material below is a brief résumé of Chapters 2 to 4 of EATS. The properties of maximum likelihood estimators are set out and this is followed by a sub-section on linear regression. It is assumed that the reader is familiar with regression and the main reason for considering it here is to introduce notation and establish a link between least squares and maximum likelihood. The prediction error decomposition is then described. This is a fundamental result in time series analysis in that it provides a method of breaking the likelihood function down into a manageable form. For univariate models, the problem of maximising the likelihood function can, by a suitable approximation, be converted into one of minimising a sum of squares function.

Maximum Likelihood

The joint density function of the observations, $y_1, \ldots, y_T$, is assumed to depend on a set of n unknown parameters in the vector $\psi = (\psi_1, \ldots, \psi_n)'$. It will be denoted by $L(y_1, \ldots, y_T; \psi)$. Once

the sample has been drawn, $L(y_1, \ldots, y_T; \psi)$ may be re-interpreted
as a likelihood function which indicates the plausibility of different
values of ψ, given the sample. Thus the likelihood is a function of ψ,
and the maximum likelihood estimator, $\tilde{\psi}$, gives the value of ψ which
maximises $L(\psi)$.

The usual method of finding the ML estimator is to differentiate
the log–likelihood function, $\log L$, with respect to each of the n
parameters, $\psi_1, \psi_2, \ldots, \psi_n$. Setting the derivatives equal to zero
yields the *likelihood equations*[1]

$$\frac{\partial \log L}{\partial \psi} = 0. \tag{2.1}$$

As a rule, the likelihood equations are non-linear and so the ML
estimates must be found by some kind of iterative procedure.

The classical theory of maximum likelihood estimation was
developed for independent and identically distributed observations.
Time series models are, by definition, concerned with observations
which are not independent. However, the standard results do apply
to time series models and the relevant formulae emerge in exactly the
same way as they do for independent observations. The main point is
that, subject to regularity conditions, $\tilde{\psi}$ is consistent and $\sqrt{T}(\tilde{\psi} - \psi)$
converges in distribution to a multivariate normal, with mean vector
zero and covariance matrix equal to the inverse of the asymptotic
information matrix. This matrix is defined by

$$IA(\psi) = \text{plim } T^{-1}\left(-\frac{\partial^2 \log L}{\partial \psi \, \partial \psi'}\right). \tag{2.2}$$

An alternative way of expressing this result is to say that $\tilde{\psi}$ is
asymptotically normal with mean ψ and covariance matrix,

$$\text{Avar}(\tilde{\psi}) = T^{-1}IA^{-1}(\psi). \tag{2.3}$$

This can be written concisely as

$$\tilde{\psi} \sim AN(\psi, T^{-1}IA^{-1}(\psi)). \tag{2.4}$$

In practice $\text{Avar}(\tilde{\psi})$ must be estimated from the sample. In order
to construct asymptotically valid test statistics, it is necessary for an
estimate, $\text{avar}(\tilde{\psi})$, to satisfy the condition

$$\text{plim } T \cdot \text{avar}(\tilde{\psi}) = IA^{-1}(\psi). \tag{2.5}$$

[1] If $f(x)$ is a function of the n variables in the vector $x = (x_1, x_2, \ldots, x_n)'$, the notation
$\partial f/\partial x$ denotes an $n \times 1$ vector in which the ith element is $\partial f/\partial x_i$. The corresponding row
vector, $(\partial f/\partial x)'$, may be written $\partial f/\partial x'$.

   The $n \times n$ matrix of second derivates, $\partial^2 \log f/\partial x \, \partial x'$, has $\partial f(x)/\partial x_i \, \partial x_j$ as its ijth
element.

If this condition is satisfied, the square roots of the diagonal elements of avar($\tilde{\psi}$) are *asymptotic standard errors*.

Linear Regression

The classical linear regression model may be written in the form

$$y_t = x_t'\beta + \epsilon_t, \qquad t = 1, \ldots, T, \tag{2.6}$$

where x_t is a $k \times 1$ vector of observations on a set of variables which can be treated as fixed in repeated samples, β is a $k \times 1$ vector of observations and ϵ_t is a random disturbance term.

If the observations are normally distributed, i.e. if $\epsilon_t \sim NID(0, \sigma^2)$, the log–likelihood function is

$$\log L(\beta, \sigma^2) = -\frac{T}{2}\log 2\pi - \frac{T}{2}\log \sigma^2 - \frac{1}{2\sigma^2}\sum_{t=1}^{T}(y_t - x_t'\beta)^2. \tag{2.7}$$

The likelihood equations are:

$$\frac{\partial \log L}{\partial \beta} = \sigma^{-2}\sum_{t=1}^{T}x_t(y_t - x_t'\beta) = 0 \tag{2.8}$$

and

$$\frac{\partial \log L}{\partial \sigma^2} = -\frac{T}{2\sigma^2} + \frac{1}{2\sigma^4}\sum_{t=1}^{T}(y_t - x_t'\beta)^2 = 0. \tag{2.9}$$

Solving (2.8) and (2.9) yields the ML estimators of β and σ^2,

$$\tilde{\beta} = \left(\sum_{t=1}^{T}x_t x_t'\right)^{-1}\sum_{t=1}^{T}x_t y_t \tag{2.10}$$

and

$$\tilde{\sigma}^2 = T^{-1}\sum_{t=1}^{T}(y_t - x_t'\tilde{\beta})^2. \tag{2.11}$$

Thus a direct solution to the likelihood equations is possible, although, as noted in the previous sub-section, this is the exception rather than the rule. The ML estimator in (2.10) is identical to the *ordinary least squares* (OLS) estimator of β, and it can be shown to be the *minimum variance unbiased estimator* (MVUE) of β. If the assumption of normality is dropped, the OLS estimator still has minimum variance within the class of linear estimators. It is said to be the *best linear unbiased estimator* (BLUE) of β.

Model (2.6) can also be written as

$$y = X\beta + \epsilon, \tag{2.12}$$

where y is the $T \times 1$ vector, $y = (y_1, \ldots, y_T)'$, X is the $T \times k$ matrix, $X = (x_1, \ldots, x_T)'$ and ϵ is the $T \times 1$ vector $\epsilon = (\epsilon_1, \ldots, \epsilon_T)'$. The OLS estimator, (2.10), is then

$$b = \tilde{\beta} = (X'X)^{-1}X'y. \tag{2.13}$$

If the disturbances have zero mean, but are serially correlated and/or have different variances, the generalised regression model is obtained. The vector ϵ in (2.12) may be replaced by a vector u which has $E(u) = 0$ and $E(uu') = \sigma^2 V$, where σ^2 is a scalar and V is a p.d. matrix. If $V = I$, the classical model is obtained and (2.13) is the BLUE of β. More generally, the BLUE of β is the *generalised least squares* (GLS) estimator,

$$\tilde{\beta} = (X'V^{-1}X)^{-1}X'V^{-1}y. \tag{2.14}$$

Again, if the disturbances are normally distributed, $\tilde{\beta}$ is the ML estimator of β.

Prediction Error Decomposition

The prediction error decomposition is a fundamental result in time series analysis, in that it enables the joint density of the observations to be written down in a manageable form. This opens the way to full maximum likelihood estimation of relatively complex time series models, as well as suggesting suitable approximations.

Consider a set of T dependent observations with mean μ and covariance matrix, $\sigma^2 V$, drawn from a multivariate normal distribution, i.e. $y \sim N(\mu, \sigma^2 V)$. The motivation behind the introduction of the scalar quantity, σ^2, is one of convenience. Its inclusion is peripheral to the main result, which would remain essentially the same if σ^2 were set equal to one. In many time series models the elements of μ will be identical and in some cases will be zero. However, allowing μ to be unrestricted permits greater generality than might seem to be the case at first sight. In particular, if $\mu = X\beta$, where X and β are as defined in (2.12), the generalised regression model is obtained.

The joint density of the observations is

$$\log L(y) = -(T/2)\log 2\pi - (T/2)\log \sigma^2 - (1/2)\log |V|$$
$$- (1/2)\sigma^{-2}(y - \mu)'V^{-1}(y - \mu). \tag{2.15}$$

This may be factored into two parts by writing

$$\log L(y) = \log L(y_1, \ldots, y_{T-1}) + \log l(y_T/y_{T-1}, \ldots, y_1).$$
$$(2.16)$$

The second term on the right-hand side of (2.16) is the distribution of the last observation, *conditional* on all the previous observations. Splitting the likelihood up in this way follows from the definition of conditional probability. In its most elementary form this states that

$$Pr(A) = Pr(A/B) \cdot Pr(B),$$

where A and B are events. Since $\log L(y)$ is, itself, a probability, namely the probability of obtaining the sample $y_1, \ldots, y_T$, equating A with y and B with the subset of y which excludes y_T gives (2.16).

Now consider the problem of estimating y_T, given that $y_{T-1}, \ldots, y_1$ are known. If $\hat{y}_{T/T-1}$ is an estimator of y_T constructed on this basis, the estimation, or prediction, error may be split into two parts:

$$y_T - \hat{y}_{T/T-1} = [y_T - E(y_T/y_{T-1}, \ldots, y_1)]$$
$$+ [E(y_T/y_{T-1}, \ldots, y_1) - \hat{y}_{T/T-1}], \qquad (2.17)$$

where $E(y_T/y_{T-1}, \ldots, y_1)$ is the mean of the distribution of y_T, conditional on $y_{T-1}, \ldots, y_1$. It follows from (2.17) that

$$\text{MSE}(\hat{y}_{T/T-1}) = \text{Var}(y_T/y_{T-1}, \ldots, y_1)$$
$$+ E\{[\hat{y}_{T/T-1} - E(y_T/y_{T-1}, \ldots, y_1)]^2\}.$$
$$(2.18)$$

The first term on the right-hand side of (2.18) is independent of $\hat{y}_{T/T-1}$, and so the *minimum mean square estimator* (MMSE) of y_T conditional on $y_{T-1}, \ldots, y_1$ is

$$\tilde{y}_{T/T-1} = E(y_T/y_{T-1}, \ldots, y_1).$$

The prediction error variance associated with $\tilde{y}_{T/T-1}$ is $E[(y_T - \tilde{y}_{T/T-1})^2]$. This is identical to $\text{MSE}(\tilde{y}_{T/T-1})$, which in view of the construction of $\tilde{y}_{T/T-1}$, is equal to $\text{Var}(y_T/y_{T-1}, \ldots, y_1)$. This quantity will be denoted by $\sigma^2 f_T$.

Since the observations are normally distributed, both components on the right-hand side of (2.16) are normal. The second term may be written

$$\log l(y_T/y_{T-1}, \ldots, y_1) = -(1/2)\log 2\pi - (1/2)\log \sigma^2$$
$$- (1/2)\log f_T - (1/2)\sigma^{-2}(y_T - \tilde{y}_{T/T-1})^2/f_T,$$

and interpreted as the distribution of the prediction error, $y_T - \tilde{y}_{T/T-1}$.

The decomposition of (2.16) may be repeated with respect to the likelihood of the first $T-1$ observations, and then repeated further until we obtain

$$\log L(y) = \sum_{t=2}^{T} \log l(y_t/y_{t-1}, \ldots, y_1) + \log l(y_1). \qquad (2.19)$$

For $t = 2, \ldots, T$, the mean of y_t conditional on $y_{t-1}, \ldots, y_1$, must be equal to $\tilde{y}_{t/t-1}$, the MMSE of y_t given the previous observations. Each of the conditional distributions is therefore the distribution of the error associated with the optimal predictor, while $l(y_1)$ is the unconditional distribution of y_1. However, if μ_1 is regarded as being the MMSE of y_1, given no previous observations, the term $y_1 - \mu_1$ is the prediction error associated with y_1. It is therefore appropriate to denote the variance of y_1 by $\sigma^2 f_1$.

Expression (2.19) allows the likelihood function to be decomposed into the joint distribution of T independent prediction errors,

$$\nu_t = y_t - \tilde{y}_{t/t-1}, \qquad t = 1, \ldots, T, \qquad (2.20)$$

where $\tilde{y}_{1/0} = \mu_1$. Each prediction error has mean zero and variance $\sigma^2 f_t$ and so (2.15) becomes

$$\log L(y) = -\frac{T}{2} \log 2\pi - \frac{T}{2} \log \sigma^2 - \frac{1}{2} \sum_{t=1}^{T} \log f_t - \frac{1}{2} \sigma^{-2} \sum_{t=1}^{T} \nu_t^2/f_t.$$

$$(2.21)$$

The prediction error decomposition may be interpreted in terms of a *Cholesky decomposition* of V^{-1}. If $\bar{L}$ is a lower triangular matrix with ones on the leading diagonal, V^{-1} may be factorised as $V^{-1} = \bar{L}'D\bar{L}$, where $D = \mathrm{diag}(f_1^{-1}, \ldots, f_T^{-1})$. This factorisation is unique and the prediction errors in (2.20) are given by the transformation $\nu = \bar{L}y$. Since the Jacobian of this transformation is $|\bar{L}| = 1$, the joint distribution of the elements in the vector $\nu = (\nu_1, \ldots, \nu_T)'$ is as in (2.21). Note that $|V^{-1}| = |\bar{L}'| \cdot |D| \cdot |\bar{L}| = |D|$ and so $\log |V| = \Sigma \log f_t$.

The Cholesky decomposition can be used as the basis for factorising V^{-1} numerically and computing the prediction errors directly from the transformation $\bar{L}y$. Alternatively, the underlying model may be cast in state space form and the prediction errors computed by the Kalman filter. This is the approach adopted in Chapter 5.

Considerable simplification can usually be achieved in the prediction error decomposition if it is felt that an approximation to the likelihood function will be adequate in practice. As a general rule, an approximation can be justified on theoretical grounds by making certain assumptions about the initial observations. This usually results in prediction errors with a constant variance, σ^2, and so $f_t = 1$ for all t. The log–likelihood function becomes

$$\log L = -\frac{T}{2}\log 2\pi - \frac{T}{2}\log \sigma^2 - \frac{1}{2}\sigma^{-2}\sum \epsilon_t^2, \qquad (2.22)$$

where the change in notation from ν_t to ϵ_t serves to indicate that each prediction error will now coincide with an underlying white noise disturbance in the model. If the likelihood function can be reduced to a form like (2.22), there are important implications for estimation. Maximising the likelihood function is equivalent to minimising the sum of squares function,

$$S(\psi) = \sum \epsilon_t^2, \qquad (2.23)$$

where ψ denotes all the parameters in the model apart from σ^2.

Example 1   The first order autoregressive model

$$y_t = \phi y_{t-1} + \epsilon_t, \qquad |\phi| < 1, \qquad (2.24)$$

where $\epsilon_t \sim NID(0, \sigma^2)$ was introduced in Section 1. In constructing the likelihood function, it will be assumed that the process began at some time in the distant past, but is only observed at times $t = 1, \ldots, T$.

For any t, the distribution of y_t, given all past observations, has a mean of ϕy_{t-1} and a variance of σ^2. Thus the last $T - 1$ prediction errors are identical to the last $T - 1$ disturbances since

$$\nu_t = y_t - \tilde{y}_{t/t-1} = y_t - \phi y_{t-1} = \epsilon_t, \qquad t = 2, \ldots, T. \quad (2.25)$$

Corresponding to (2.19), the log–likelihood function may be decomposed into the density functions of $\epsilon_2, \ldots, \epsilon_T$, plus the unconditional density of y_1. This is normally distributed with mean 0 and variance $\sigma^2/(1 - \phi^2)$; see (2.2.7). The first prediction error is therefore equal to y_1 itself, rather than being equal to the disturbance ϵ_1. Setting $\nu_1 = y_1$, the log–likelihood function for all T observations may be written in the form (2.21), i.e.

$$\log L(y) = -\frac{T}{2}\log 2\pi - \frac{T}{2}\log \sigma^2 + \frac{1}{2}\log(1 - \phi^2)$$

$$-\frac{1}{2}\sigma^{-2}(1 - \phi^2)y_1^2 - \frac{1}{2}\sigma^{-2}\sum_{t=2}^{T}(y_t - \phi y_{t-1})^2$$

$$(2.26)$$

Note that $f_t = 1$ for $t = 2, \ldots, T$, while $f_1 = (1 - \phi^2)^{-1}$.

The third and fourth terms in (2.26) are asymptotically negligible, and if they are removed the likelihood function simplifies to the form (2.22). A formal justification for doing this is to regard y_1 as being fixed. The ML estimator of ϕ is then linear, being given by a regression of y_t on y_{t-1}.

Decomposition of the Likelihood Function for Multivariate Models

The prediction error decomposition may be extended to multivariate series where an $N \times 1$ vector is observed at each point in time. The argument goes through exactly as before, with ν_t being an $N \times 1$ vector of prediction errors at time t, with mean zero and covariance matrix F_t. (There is less point in extracting a scalar quantity, σ^2, in the multivariate case.) The log–likelihood may be decomposed as

$$\log L(y_1, \ldots, y_T) = -\frac{TN}{2} \log 2\pi - \frac{1}{2} \sum_{t=1}^{T} \log |F_t|$$

$$-\frac{1}{2} \sum_{t=1}^{T} \nu_t' F_t^{-1} \nu_t, \tag{2.27}$$

where $y_1 \sim N(\mu_1, F_1)$ and $\nu_1 = y_1 - \mu_1$.

The Kalman filter can again be used to effect the decomposition in (2.27), provided the underlying model can be expressed in state space form. However, as in the univariate case, it is often possible to approximate $\log L$ by identifying ν_t with the vector of white noise disturbances driving the process. If $\epsilon_t \sim NID(0, \Omega)$, the approximation takes the form

$$\log L = -\frac{TN}{2} \log 2\pi - \frac{T}{2} \log |\Omega| - \frac{1}{2} \sum \epsilon_t' \Omega^{-1} \epsilon_t. \tag{2.28}$$

Maximisation of a Likelihood Function

There are a number of different approaches to numerical optimisation. However, when the problem is to maximise a likelihood function, an important class of algorithms is defined by iterations of the form

$$\psi^* = \hat{\psi} + [I^*(\hat{\psi})]^{-1} \{\partial \log L / \partial \psi\}. \tag{2.29}$$

$I^*(\psi)$ denotes the information matrix, or an approximation to it, and in (2.29) this is evaluated at the current estimate, $\hat{\psi}$. The vector of derivatives, $\partial \log L / \partial \psi$, is also evaluated at $\hat{\psi}$ and application of (2.29) yields an updated estimate ψ^*. This process is repeated until convergence. In practice, some modification of the scheme will usually be made, for example by the introduction of a variable step length; see EATS (Section 4.3).

When $I^*(\psi)$ is defined as the Hessian of $-\log L$, the iterative procedure in (2.29) is known as *Newton–Raphson*. Taking the expectation of the Hessian yields the information matrix itself and the scheme is then known as the *method of scoring*. In the special case when maximising the likelihood function is equivalent to minimising a sum of squares function, (2.23),

$$\frac{\partial \log L}{\partial \psi} = \frac{1}{2\sigma^2} \sum z_t \epsilon_t,$$

where

$$z_t = -\partial \epsilon_t / \partial \psi.$$

Defining $I^*(\psi)$ as

$$I^*(\psi) = \sigma^{-2} \sum z_t z_t',$$

leads to the iterative scheme

$$\psi^* = \hat{\psi} + (\sum z_t z_t')^{-1} \sum z_t \epsilon_t. \tag{2.30}$$

This is known as *Gauss–Newton*. At each step, ϵ_t and its vector of derivatives, z_t, are evaluated at the current estimate, $\hat{\psi}$. This estimate is then 'corrected' by regressing ϵ_t on z_t. The iterative scheme is completely independent of the parameter σ^2. Once the procedure has converged, this can be estimated from the residuals, $\epsilon_t(\tilde{\psi})$, using the formula,

$$\tilde{\sigma}^2 = T^{-1} \sum \epsilon_t^2(\tilde{\psi}). \tag{2.31}$$

Two-Step Estimators

Equation (2.29) can be used as the basis for deriving asymptotically efficient *two-step* estimators. If $\hat{\psi}$ is a *consistent* estimator of ψ and certain regularity conditions are satisfied, ψ^* will have the same asymptotic distribution as the ML estimator; cf. (2.4). However, its small sample properties may be very different.

3. Testing

Methods for testing hypotheses can be derived systematically using a maximum likelihood approach. The basic test procedure is the likelihood ratio test, but two other tests, the Wald test and Lagrange multiplier test, have similar large sample properties and are often more attractive. All three tests are described in the first sub-section. The second sub-section goes on to deal with certain aspects of testing which do not fit neatly into this classical framework. A more detailed discussion will be found in Chapter 5 of EATS.

Classical Test Procedures

The likelihood ratio (LR) test is primarily concerned with testing the validity of a set of restrictions on the parameter vector, ψ. When these restrictions are linear, they may be expressed in the form

$$R\psi = r, \tag{3.1}$$

where R is an $m \times n$ matrix of fixed values, r is an $m \times 1$ vector of fixed values and m, the number of restrictions, is less than n.

Under the null hypothesis, H_0, ψ satisfies the restrictions in (3.1). When the restrictions are imposed, the ML estimator of ψ is denoted by $\tilde{\psi}_0$ and this may be contrasted with the unrestricted estimator, $\tilde{\psi}$. If the maximised likelihood function under H_0, $L(\tilde{\psi}_0)$, is much smaller than the unrestricted maximised likelihood, $L(\tilde{\psi})$, there is evidence against the null hypothesis. This result is formalised in the Neyman–Pearson lemma which shows that a test based on the likelihood ratio,

$$\lambda = L(\tilde{\psi}_0)/L(\tilde{\psi}), \tag{3.2}$$

has certain desirable statistical properties.

It is sometimes possible to transform the likelihood ratio into a statistic, the exact distribution of which is known under H_0. When this cannot be done, a large sample test is carried out. This is based on the result that the statistic

$$\text{LR} = -2 \log \lambda \tag{3.3}$$

is asymptotically distributed as χ_m^2 under H_0.

The disadvantage of the LR test is that the model must be estimated under both the null and alternative hypotheses. A different procedure, the *Wald test*, only requires an estimate of ψ from the unrestricted model. The usual form of the test statistic is

$$W = (R\tilde{\psi} - r)'[RI^{-1}(\tilde{\psi})R']^{-1}(R\tilde{\psi} - r), \tag{3.4}$$

where $I(\tilde{\psi})$ is the information matrix evaluated at the unrestricted estimate $\tilde{\psi}$. Under H_0, W is asymptotically χ_m^2 and its large sample properties can be shown to be similar to those of the LR test.

If the model is easier to estimate under the null hypothesis, a *Lagrange multiplier* test may be appropriate. The test statistic, which again is asymptotically χ_m^2 under H_0, takes the form

$$\text{LM} = \left(\frac{\partial \log L}{\partial \psi}\right)' I^{-1}(\tilde{\psi}_0)\left(\frac{\partial \log L}{\partial \psi}\right), \tag{3.5}$$

where $\partial \log L/\partial \psi$ is evaluated at the restricted estimate, $\tilde{\psi}_0$. As with the Wald test, the large sample properties of the LM test are similar to those of the LR test, but estimation of the more general, unrestricted model is avoided.

When maximising the likelihood function of the unrestricted model is equivalent to minimising a sum of squares function, a minor modification of (3.5) leads to a particularly convenient form of the test statistic. If the residual, ϵ_t, and its $m \times 1$ vector of derivatives, $\partial \epsilon_t/\partial \psi$, are evaluated at $\psi = \tilde{\psi}_0$, ϵ_t may be regressed on $\partial \epsilon_t/\partial \psi$ to yield a coefficient of multiple correlation, R^2. The statistic

$$\text{LM*} = TR^2 \tag{3.6}$$

is then asymptotically equivalent to the LM statistic, (3.5), and it can be tested as a χ_m^2 variate in the usual way. On occasion it is convenient to adopt a 'modified LM test' based on the F-distribution. The general principle underlying this approach should be clear from the examples given.

Non-Nested Hypotheses and Diagnostic Checks

The classical test procedures are concerned with nested hypotheses, the model under the null being a special case of the more general model. Sometimes, however, it is necessary to compare two models, neither of which is a special case of the other. Such models are termed *non-nested* and there are basically two ways of treating them. The first is to define some criterion for discriminating between them. The second is to employ a test procedure which tests each formulation against the other. A procedure of the second kind has four possible outcomes, since in addition to one model being accepted and the other rejected, it is also possible for both to be accepted or both to be rejected.

The maximised likelihood function provides a possible criterion for discriminating between models. However, it is advisable to make some allowance for the number of parameters estimated. This is formalised in the *Akaike Information Criterion* (AIC), where the decision rule is to select the model for which

$$\text{AIC} = -2 \log L(\tilde{\psi}) + 2n \tag{3.7}$$

is a minimum; n is the number of parameters. Methods of testing one model against another are rather more complicated and are not used in this book.

Diagnostic checking is another aspect of testing which does not fit neatly into the classical framework. If a model is adequate, the residuals should be approximately random and most diagnostic checking procedures represent attempts to detect departures from randomness. The procedures may be graphical or they may take the form of formal test statistics, but the essential feature of diagnostic checks is that the researcher has no firm views as to possible sources of model inadequacy. Thus, although test statistics for diagnostic checking can be derived with particular alternatives in mind, say by the LM approach, a rejection of the null hypothesis need not necessarily point in the direction of that particular alternative. A classic example is the Durbin–Watson test in linear regression. This test was derived to have high power against the alternative of a first-order autoregressive disturbance term. However, a significant value of the test statistic can be an indication of a wide range of possible misspecifications, including incorrect functional form or the omission of a key variable.

Exercises

1. The following values are for US Personal Consumption Expenditures – Household Operation (Billions of Current Dollars) for the years 1967–71: 29.2, 31.2, 33.8, 36.4, 39.4. Use the global linear trend method (i.e. linear regression) to estimate a slope and to give a forecast of the value for 1972.

In fact the values for 1972–4 were: 43.3, 47.3, 52.9. Use the local linear trend method (i.e. the Holt–Winters method), with smoothing constants equal to 0.1, to update the means and slopes for each year starting from the equation fitted in the first part of the question, and thereby give a forecast of the value for 1975. Comment on the adequacy of the method in dealing with this data. [University of York, 1980]

2. Show that the estimator of μ obtained by minimising the discounted sum of squares,

$$S(\mu) = \sum_{t=1}^{T} \delta^{T-t}(y_t - \mu)^2, \qquad 0 < \delta \leqslant 1,$$

with respect to μ is a weighted average of observations in which the observations have exponentially declining weights. If this estimator is used as a forecast, $\tilde{y}_{T+1/T}$, of y_{T+1}, show that this forecast satisfies the recursion

$$\tilde{y}_{T+1/T} = (1 - \delta)y_T + \delta \tilde{y}_{T/T-1},$$

where $\tilde{y}_{T/T-1}$ is the forecast of y_T based on the first $T - 1$ observations. Comment on the relationship between the value of δ and the concept of local and global trends.

2
Stationary Stochastic Processes and their Properties in the Time Domain

1. Basic Concepts

This book is primarily concerned with modelling time series as stochastic processes. Each observation in a stochastic process is a random variable, and the observations evolve in time according to certain probabilistic laws. Thus a stochastic process may be defined as a collection of random variables which are ordered in time.

The model defines the mechanism by which the observations are generated. A simple example is the first-order moving average process,

$$y_t = \epsilon_t + \theta \epsilon_{t-1}, \qquad t = 1, \ldots, T, \tag{1.1}$$

where ϵ_t is a sequence of uncorrelated random variables with mean zero and constant variance and θ is a parameter. A particular set of values of $\epsilon_0, \epsilon_1, \ldots, \epsilon_T$, results in a corresponding sequence of observations, $y_1, \ldots, y_T$. By drawing a different set of values of $\epsilon_0, \epsilon_1, \ldots, \epsilon_T$, a different set of observations is obtained and model (1.1) can be regarded as being capable of generating an infinite set of such *realisations* over the period $t = 1, \ldots, T$. Thus, the model effectively defines a joint distribution for the random variables $y_1, \ldots, y_T$.

The moments of a stochastic process are defined with respect to the distribution of the random variables $y_1, \ldots, y_T$. The mean of the process at time t is

$$\mu_t = E(y_t), \qquad t = 1, \ldots, T, \tag{1.2}$$

and this can be interpreted as the average value of y_t taken over all possible realisations. Second moments have a similar interpretation. The variance at time t is defined by

$$\mathrm{Var}(y_t) = E[(y_t - \mu_t)^2], \qquad t = 1, \ldots, T, \tag{1.3}$$

while the covariance between y_t and $y_{t-\tau}$ is given by

$$\mathrm{Cov}(y_t, y_{t-\tau}) = E[(y_t - \mu_t)(y_{t-\tau} - \mu_{t-\tau})], \qquad t = \tau + 1, \ldots, T, \tag{1.4}$$

If several realisations are available, the above quantities can be sensibly estimated by 'ensemble' averages. For example,

$$\hat{\mu}_t = m^{-1} \sum_{j=1}^{m} y_t^{(j)}, \qquad t = 1, \ldots, T, \tag{1.5}$$

where $y_t^{(j)}$ denotes the jth observation on y_t and m is the number of realisations. However, in most time series problems, only a single series of observations is available. In these circumstances, no meaningful inferences can be made about the quantities defined in (1.2) to (1.4), unless some restrictions are placed on the process which is assumed to be generating the observations. This leads to the concept of stationarity.

Stationarity

When only a single realisation of observations is available, attention must shift from the aggregation of observations at particular points *in* time, to the averaging of observations *over* time. This is only possible if the data generating process is such that the quantities (1.2), (1.3) and (1.4) are independent of time. Thus, for example, if $\mu_t = \mu$ for $t = 1, \ldots, T$, it can be estimated by taking the average of the observations $y_1, \ldots, y_T$.

For a stochastic process to be *stationary*, the following conditions must be satisfied for all values of t:

$$E(y_t) = \mu, \tag{1.6}$$

$$E[(y_t - \mu)^2] = \gamma(0), \tag{1.7}$$

and

$$E[(y_t - \mu)(y_{t-\tau} - \mu)] = \gamma(\tau), \qquad \tau = 1, 2, \ldots \tag{1.8}$$

Expressions (1.6) and (1.7) define the mean and variance of the data generating process, while (1.8) gives the autocovariance at lag τ. The implications of (1.6) and (1.7) were noted in the discussion surrounding figure 1.1(b).

The quantities (1.6) to (1.8) can be estimated from a single series of observations as follows:

$$\hat{\mu} = \bar{y} = T^{-1} \sum_{t=1}^{T} y_t, \tag{1.9}$$

$$\hat{\gamma}(0) = c(0) = T^{-1} \sum_{t=1}^{T} (y_t - \bar{y})^2, \tag{1.10}$$

and

$$\hat{\gamma}(\tau) = c(\tau) = T^{-1} \sum_{t=\tau+1}^{T} (y_t - \bar{y})(y_{t-\tau} - \bar{y}), \qquad \tau = 1, 2, 3, \ldots \tag{1.11}$$

If the process is *ergodic*, these statistics give consistent estimates of the mean, variance and autocovariances. Ergodicity will not be defined formally here, but what it basically requires is that observations sufficiently far apart should be almost uncorrelated. For all the models considered in this chapter, stationarity implies ergodicity and it is unnecessary to dwell on the matter any further.

The conditions (1.6) to (1.8) provide a definition of *weak* or *covariance* stationarity. Occasionally the condition of *strict* stationarity is imposed. This is a stronger condition whereby the joint probability of a set of r observations at times $t_1, t_2, \ldots, t_r$ is the same as the joint probability of the observations at times $t_1 + \tau, t_2 + \tau, \ldots, t_r + \tau$. Here, the term stationarity will always refer to weak stationarity, though it should be noted that if a series is weakly stationary and normally distributed, then it is also stationary in the strict sense.

The simplest example of a stationary stochastic process is a sequence of uncorrelated random variables with constant mean and variance. A process of this kind is known as *'white noise'*, a terminology borrowed from the engineering literature. Throughout this book, the symbol ϵ_t will always denote a white noise variable, and unless explicitly stated otherwise, such a variable will have a mean of zero and a variance of σ^2. Because the variables in a white noise sequence are uncorrelated, the autocovariances at non-zero lags are all zero. Thus

$$E(\epsilon_t \epsilon_{t-\tau}) = \begin{cases} \sigma^2, & \tau = 0 \\ 0, & \tau \neq 0. \end{cases} \tag{1.12}$$

Autocovariance and Autocorrelation Functions

When a stochastic process is stationary, its time domain properties can be summarised by plotting $\gamma(\tau)$ against τ. This is known as the *autocovariance function.* Since $\gamma(\tau) = \gamma(-\tau)$, it is unnecessary to extend the plot over negative values of τ.

The autocovariances may be standardised by dividing through by the variance of the process. This yields the autocorrelations,

$$\rho(\tau) = \gamma(\tau)/\gamma(0), \qquad \tau = 0, \pm 1, \pm 2, \ldots \tag{1.13}$$

A plot of $\rho(\tau)$ against non-negative values of τ gives the *auto-correlation function*. Note that $\rho(0) = 1$ by definition.

The time domain properties of the moving average model, (1.1), are relatively easy to derive. Since ϵ_t is a white noise variable with mean 0 and variance σ^2, it follows that

$$\mu = E(\epsilon_t) + \theta E(\epsilon_{t-1}) = 0, \tag{1.14a}$$

while

$$\begin{aligned}
\gamma(0) &= E[(\epsilon_t + \theta\epsilon_{t-1})(\epsilon_t + \theta\epsilon_{t-1})] \\
&= E(\epsilon_t^2) + \theta^2 E(\epsilon_{t-1}^2) + 2\theta E(\epsilon_t\epsilon_{t-1}) \\
&= (1 + \theta^2)\sigma^2.
\end{aligned} \tag{1.14b}$$

Similarly,

$$\begin{aligned}
\gamma(1) &= E[(\epsilon_t + \theta\epsilon_{t-1})(\epsilon_{t-1} + \theta\epsilon_{t-2})] \\
&= E(\epsilon_t\epsilon_{t-1}) + \theta E(\epsilon_{t-1}^2) + \theta E(\epsilon_t\epsilon_{t-2}) + \theta^2 E(\epsilon_{t-1}\epsilon_{t-2}) \\
&= \theta E(\epsilon_{t-1}^2) \\
&= \theta\sigma^2,
\end{aligned} \tag{1.14c}$$

and

$$\gamma(\tau) = 0, \qquad \tau = 2, 3, 4, \ldots \tag{1.14d}$$

The mean, variance and covariances are therefore independent of t and the process is stationary.

The autocovariance and autocorrelation functions have exactly the same shape and provide the same information on the nature of the process. It is more usual to plot the autocorrelation function since it is dimensionless. Standardising (1.14c) gives

$$\rho(1) = \theta/(1 + \theta^2) \tag{1.15}$$

and the autocorrelation function for $\theta = 0.5$ is shown in figure 2.1.

When θ is positive, successive values of y_t are positively correlated and so the process will tend to be smoother than the random series, ϵ_t. On the other hand, a negative value of θ will yield a series which is more irregular than a random series, in the sense that positive values of y_t tend to be followed by negative values and *vice*

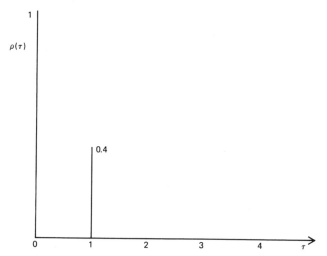

Figure 2.1   *Autocorrelation Function of an MA(1) Process with θ = 0.5*

versa. This is reflected in the autocorrelation function, as $\rho(1)$ is negative for $\theta < 0$.

The Correlogram

The quantities defined in (1.9), (1.10) and (1.11) are the *sample mean, sample variance* and *sample autocovariances* respectively. The sample autocovariances may be standardised in the same way as theoretical autocovariances. This yields the sample autocorrelations,

$$r(\tau) = c(\tau)/c(0), \qquad \tau = 1, 2, \ldots, \tag{1.16}$$

and a plot of $r(\tau)$ against non-negative values of τ is known as the sample autocorrelation function or *correlogram*.

The sample autocorrelations are estimates of the corresponding theoretical autocorrelations for the stochastic process which is assumed to be generating the data. They will therefore be subject to sampling variability, and so although the correlogram will tend to mirror the properties of the theoretical autocorrelation function, it will not reproduce them exactly. The sample autocorrelations from a white noise process, for example, will typically be close to zero but will not, in general, be identically equal to zero.

The correlogram is the main tool for analysing the properties of a series of observations in the time domain. However, in order to interpret the correlogram it is necessary to know something, firstly about the sampling variability of the estimated autocorrelations, and secondly about the autocovariance functions of different stochastic processes. The sections which follow examine the nature of the autocorrelation function for various special cases within the class of autoregressive-moving average models. This provides the basis for the model building strategy developed in Chapter 6. The question of sampling variability, together with the related question of test procedures associated with the correlogram, is also taken up in that chapter.

The Lag Operator

The lag operator, L, plays an extremely useful role in carrying out algebraic manipulations in time series analysis. It is defined by the transformation

$$Ly_t = y_{t-1}. \tag{1.17}$$

Applying L to y_{t-1} yields $Ly_{t-1} = y_{t-2}$. Substituting from (1.17) gives $L(Ly_t) = L^2 y_t = y_{t-2}$ and so, in general,

$$L^\tau y_t = y_{t-\tau}, \qquad \tau = 1, 2, 3, \ldots \tag{1.18}$$

It is logical to complete the definition by letting L^0 have the property $L^0 y_t = y_t$ so that (1.18) holds for all non-negative integers.

The lag operator can be manipulated in a similar way to any algebraic quantity. Consider an infinite moving average process in which the coefficient of ϵ_{t-j} is ϕ^j for $j = 0, 1, 2, \ldots$, and $|\phi| < 1$. The model may be written

$$y_t = \sum_{j=0}^{\infty} (\phi L)^j \epsilon_t \tag{1.19}$$

and if L is regarded as having the property $|L| \leq 1$, it follows that $|\phi L| < 1$ and so the series $1, \phi L, (\phi L)^2, \ldots$ may be summed as an infinite geometric progression. Thus (1.19) becomes

$$y_t = \epsilon_t/(1 - \phi L) \tag{1.20}$$

and this may be re-arranged to give the first-order autoregressive process

$$y_t = \phi y_{t-1} + \epsilon_t. \tag{1.21}$$

The *first difference operator*, Δ, can be manipulated in a similar way to the lag operator, since $\Delta = 1 - L$. The relationship between the two operators can often be usefully exploited. For example,

$$\Delta^2 y_t = (1 - L)^2 y_t = (1 - 2L + L^2)y_t = y_t - 2y_{t-1} + y_{t-2}. \tag{1.22}$$

Autoregressive-Moving Average Processes

A general class of stochastic processes can be formed by introducing an infinite number of lags into a moving average. This yields the representation

$$y_t = \sum_{j=0}^{\infty} \psi_j \epsilon_{t-j},$$ (1.23)

where $\psi_0, \psi_1, \psi_2, \ldots$ are parameters. The condition

$$\sum_{j=1}^{\infty} \psi_j^2 < \infty,$$ (1.24)

must be imposed in order to ensure that the process has finite variance. Any model which can be written in this form is said to be an *indeterministic* or *linear* process.

A linear process is stationary and its properties may be expressed in terms of the autocovariance function. These properties can be approximated, to any desired level of accuracy, by a model drawn from the class of autoregressive-moving average processes. An autoregressive-moving average process of order (p, q), which is normally abbreviated as ARMA(p, q), is written as

$$y_t = \phi_1 y_{t-1} + \cdots + \phi_p y_{t-p} + \epsilon_t + \theta_1 \epsilon_{t-1} + \cdots + \theta_q \epsilon_{t-q}.$$ (1.25)

As will be shown in the next section, certain restrictions must be placed on the autoregressive parameters, $\phi_1, \ldots, \phi_p$, if the model is to be stationary.

An ARMA process can be written more concisely by defining *associated polynomials* in the lag operator. If

$$\phi(L) = 1 - \phi_1 L - \cdots - \phi_p L^p,$$ (1.26a)

and

$$\theta(L) = 1 + \theta_1 L + \cdots + \theta_q L^q,$$ (1.26b)

model (1.25) becomes

$$\phi(L)y_t = \theta(L)\epsilon_t.$$ (1.27)

This representation has certain technical, as well as notational, advantages. For example, if a non-zero mean, μ, is introduced into a stationary model it becomes

$$y_t = \mu + \phi^{-1}(L)\theta(L)\epsilon_t.$$ (1.28)

On multiplying through by $\phi(L)$, it is apparent that the same effect could be achieved by adding a parameter

$$\theta_0 = \phi(L)\mu = \phi(1)\mu = (1 - \phi_1 - \cdots - \phi_p)\mu$$

to the right-hand side of (1.27).

The next three sections explore the properties of pure auto-regressive, pure moving average and mixed (i.e. ARMA) processes. It is assumed throughout that the processes have zero mean, i.e. $\mu = 0$. This is purely for convenience and implies no loss in generality.

2. Autoregressive Processes

An autoregressive process of order p is written as

$$y_t = \phi_1 y_{t-1} + \cdots + \phi_p y_{t-p} + \epsilon_t, \qquad t = 1, \ldots, T. \qquad (2.1)$$

This will be denoted by writing $y_t \sim AR(p)$. Autoregressive processes have always been popular, partly because they have a natural interpretation, and partly because they are easier to estimate than MA or mixed processes.

The first point to establish about an AR model is the conditions under which it is stationary. This amounts to determining whether it can be written in the form (1.23) with condition (1.24) holding. Once this has been done, the autocorrelation function can be derived.

Stationarity for the First-Order Model

The AR(1) process is

$$y_t = \phi y_{t-1} + \epsilon_t, \qquad t = 1, \ldots, T. \qquad (2.2)$$

Although the series is first observed at time $t = 1$, the process is regarded as having started at some time in the remote past. Substituting repeatedly for lagged values of y_t gives

$$y_t = \sum_{j=0}^{J-1} \phi^j \epsilon_{t-j} + \phi^J y_{t-J}. \qquad (2.3)$$

The right-hand side of (2.3) consists of two parts, the first of which is a moving average of lagged values of the white noise variable driving the process. The second part depends on the value of y_t at time $t - J$. Taking expectations and treating y_{t-J} as a fixed number yields

$$E(y_t) = E\left(\sum_{j=0}^{J-1} \phi^j \epsilon_{t-j}\right) + E(\phi^J y_{t-J}) = \phi^J y_{t-J}. \qquad (2.4)$$

If $|\phi| \geqslant 1$, the mean value of the process depends on the starting value, y_{t-J}. Expression (2.3) therefore contains a deterministic component and a knowledge of y_{t-J} enables non-trivial predictions to be made for future values of the series, no matter how far ahead. If, on the other hand, ϕ is less than one in absolute value, this deterministic component is negligible if J is large. As $J \to \infty$, it effectively disappears and so if the process is regarded as having started at some point in the remote past, it is quite legitimate to write (2.2) in the form

$$y_t = \sum_{j=0}^{\infty} \phi^j \epsilon_{t-j}, \qquad t = 1, \ldots, T. \tag{2.5}$$

On comparing (2.5) with (1.23), it can be seen that an AR(1) process with $|\phi| < 1$ is indeterministic, since summing the squared coefficients as a geometric progression yields

$$\sum_{j=0}^{\infty} \phi^{2j} = 1/(1 - \phi^2). \tag{2.6}$$

The expectation of y_t is zero for all t, while

$$\gamma(0) = E(y_t^2) = \left(\sum_{j=0}^{\infty} \phi^j \epsilon_{t-j} \right)^2 = \sum_{j=0}^{\infty} \phi^{2j} E(\epsilon_{t-j}^2)$$

$$= \sigma^2 \sum_{j=0}^{\infty} \phi^{2j} = \sigma^2/(1 - \phi^2). \tag{2.7}$$

Stationarity for the Second-Order Model*

The second-order autoregressive process is defined by

$$y_t = \phi_1 y_{t-1} + \phi_2 y_{t-2} + \epsilon_t, \qquad t = 1, \ldots, T. \tag{2.8}$$

As with the first-order model, it is possible to decompose (2.8) into two parts, one stochastic and the other deterministic. The deterministic part depends on a pair of starting values, but if the process is stationary, their influence is negligible for a starting point some time in the remote past.

In order to study the nature of the deterministic component, it is easiest to suppress the disturbance term in (2.8). This yields the homogenous difference equation,

$$\bar{y}_t - \phi_1 \bar{y}_{t-1} - \phi_2 \bar{y}_{t-2} = 0, \tag{2.9}$$

---

*Indicates a more difficult section which may be omitted without loss of continuity.

where the bar over y_t indicates that we are now dealing with the mean of the process. The solution to (2.9) depends on the roots of the characteristic equation,

$$x^2 - \phi_1 x - \phi_2 = 0; \qquad (2.10)$$

see, for example, Goldberg (1958). Since (2.10) is a quadratic equation, these roots, m_1 and m_2, satisfy

$$(x - m_1)(x - m_2) = 0, \qquad (2.11)$$

and they may be found in the usual way from the formula

$$m_1, m_2 = (\phi_1 \pm \sqrt{\phi_1^2 + 4\phi_2})/2. \qquad (2.12)$$

Three possible cases arise with regard to the solution of (2.12) depending on whether the term under the square root is positive, zero or negative. In the first case, the roots are both real and the solution to (2.9) is given by

$$\bar{y}_t = k_1 m_1^J + k_2 m_2^J, \qquad (2.13)$$

where k_1 and k_2 are constants which depend on the starting values, $\bar{y}_{t-J}$ and $\bar{y}_{t-J+1}$. If both m_1 and m_2 are less than unity in absolute value, $\bar{y}_t$ will be close to zero if J is large.

When $\phi_1^2 + 4\phi_2 < 0$, the roots are a pair of complex conjugates. The solution is again of the form (2.13), but it may be re-written as

$$\bar{y}_t = k_3 r^J \cos(\lambda J + k_4), \qquad (2.14)$$

where k_3 and k_4 are constants which depend on the starting values of the series, r is the modulus of the roots and λ is defined by

$$\lambda = \tan^{-1}[Im(m_1)/Re(m_1)] = \tan^{-1}[(-\phi_1^2 - 4\phi_2)/\phi_1]$$
$$= \cos^{-1}[|\phi_1|/(2\sqrt{-\phi_2})] \qquad (2.15)$$

and measured in radians. The time path followed by $\bar{y}_t$ is cyclical, but if the modulus of the roots is less than unity it is damped and $\bar{y}_t$ is negligible if J is large.

When the roots are real and equal, the solution to (2.9) takes a slightly different form from (2.13), but the condition that this root be less than unity is necessary for $\bar{y}_t$ to be negligible.

If J is allowed to tend to infinity and the roots of (2.10) are less than one in absolute value, the deterministic component in (2.8) disappears completely. This leaves a linear process, (1.23). The coefficients in this process can be derived most easily by defining the polynomials

$$\phi(L) = 1 - \phi_1 L - \phi_2 L^2 \qquad (2.16)$$

and

$$\psi(L) = \psi_0 + \psi_1 L + \psi_2 L^2 + \cdots + \psi_\tau L^\tau + \cdots \qquad (2.17)$$

and noting that (2.8) and (1.23) can be written as

$$y_t = \phi^{-1}(L)\epsilon_t \qquad (2.18)$$

and

$$y_t = \psi(L)\epsilon_t \qquad (2.19)$$

respectively. On comparing (2.18) and (2.19) it will be seen that

$$\phi(L)\psi(L) = 1. \qquad (2.20)$$

This can be expanded to yield

$$(1 - \phi_1 L - \phi_2 L^2)(\psi_0 + \psi_1 L + \psi_2 L^2 + \cdots) = 1$$

which on re-arrangement becomes:

$$\psi_0 + (\psi_1 - \phi_1)L + (\psi_2 - \phi_1\psi_1 - \phi_2\psi_0)L^2$$
$$+ (\psi_3 - \phi_1\psi_2 - \phi_2\psi_1)L^3 + \cdots = 1. \qquad (2.21)$$

The coefficients of $L, L^2, L^3 \ldots$ on the right-hand side of (2.21) are all zero and so

$$\psi_1 - \phi_1 = 0$$
$$\psi_j - \phi_1\psi_{j-1} - \phi_2\psi_{j-2} = 0, \qquad j \geqslant 2. \qquad (2.22)$$

For $j \geqslant 2$, ψ_j is determined by the second order difference equation, (2.22), with starting values $\psi_0 = 1$ and $\psi_1 = \phi_1$. This difference equation has exactly the same form as (2.9). Its roots are given by (2.12) and if they are both less than one in absolute value, ψ_j will tend towards zero as $j \to \infty$. It can be shown that this movement towards zero takes place quickly enough for (1.24) to be satisfied and that the condition that the roots of (2.10) have modulus less than unity is sufficient to ensure stationarity.

The conditions for stationarity may be defined in terms of the parameters ϕ_1 and ϕ_2 as follows:

$$\phi_1 + \phi_2 < 1, \qquad (2.23a)$$
$$-\phi_1 + \phi_2 < 1, \qquad (2.23b)$$
$$\phi_2 > -1; \qquad (2.23c)$$

see, for example, Goldberg (1958, pp. 171–2). The roots of the homogeneous equation will be complex if

$$\phi_1^2 + 4\phi_2 < 0, \qquad (2.24)$$

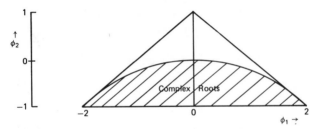

Figure 2.2   *Admissible Region for ϕ_1 and ϕ_2 in a Stationary AR(2) Process*

although a necessary condition for complex roots is simply that
$\phi_2 < 0$. These conditions are conveniently summarised in figure 2.2,
where all the points which actually lie *within* the triangle (but not
on the boundary) correspond to values of ϕ_1 and ϕ_2 in a stationary
process. The shaded area indicates complex roots.

Stationarity for the pth Order Model

The AR(p) model, (2.1), will be stationary if the roots of the
characteristic equation

$$x^p - \phi_1 x^{p-1} - \cdots - \phi_p = 0 \tag{2.25}$$

are less than one in absolute value, i.e. if they lie within the unit
circle. An alternative way of expressing this condition is in terms of
the associated polynomial, (1.26a). The polynomial equation,

$$1 - \phi_1 L - \cdots - \phi_p L^p = 0, \tag{2.26}$$

is similar in form to (2.25) except that x is replaced by $1/L$ and the
whole equation is multiplied through by L^p. The stationarity
condition is that the roots of (2.26) should be *outside* the unit
circle.

Autocovariance and Autocorrelation Functions

When $|\phi| < 1$, the AR(1) model, (2.2), has a mean of zero and a
variance given by (2.7). The autocovariance at lag τ can be derived
by expressing y_t as a linear combination of $y_{t-\tau}$ and $\epsilon_t, \epsilon_{t-1}, \ldots,$
$\epsilon_{t-\tau+1}$. This is achieved by setting $J = \tau$ in (2.3) and so

$$\gamma(\tau) = E(y_t y_{t-\tau}) = E\left[\left(\phi^\tau y_{t-\tau} + \sum_{j=0}^{\tau-1} \phi^j \epsilon_{t-j}\right) y_{t-\tau}\right].$$

Since $\epsilon_t, \ldots, \epsilon_{t-\tau+1}$ are all uncorrelated with $y_{t-\tau}$, this expression
reduces to

$$\gamma(\tau) = \phi^\tau E(y_{t-\tau}^2) = \phi^\tau \gamma(0), \qquad \tau = 1, 2, \ldots \tag{2.27}$$

The autocovariances depend only on τ, confirming that the process is stationary.

The autocovariances may be derived by a more direct method if stationarity is assumed from the outset. Multiplying both sides of (2.2) by $y_{t-\tau}$ and taking expectations gives

$$E(y_t y_{t-\tau}) = \phi E(y_{t-1} y_{t-\tau}) + E(\epsilon_t y_{t-\tau}), \qquad \tau = 0, 1, 2, \ldots \tag{2.28}$$

For a stationary process, $E(y_{t-1} y_{t-\tau}) = E(y_t y_{t-\tau+1}) = \gamma(\tau - 1)$ and, if $\tau > 0$, the last term is zero, as ϵ_t is uncorrelated with past values of y_t. Therefore

$$\gamma(\tau) = \phi \gamma(\tau - 1), \qquad \tau = 1, 2, \ldots \tag{2.29}$$

This is a first-order difference equation with a solution given by (2.27). The expression for the variance of the process may be derived by a similar method by noting that $E(\epsilon_t y_t) = \sigma^2$.

The autocorrelation function takes the form

$$\rho(\tau) = \phi^\tau, \qquad \tau = 0, 1, 2, \ldots \tag{2.30}$$

and for positive values of ϕ this exhibits a smooth exponential decay as shown in figure 2.3(a). When ϕ is negative, the autocorrelation function again decays exponentially, but it oscillates between negative and positive values as shown in figure 2.3(b). When ϕ is positive, the series is 'slowly changing' in that the differences between

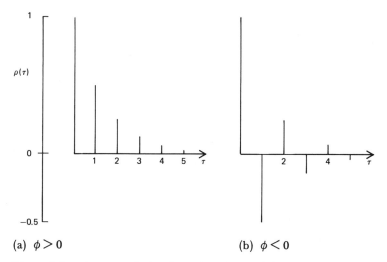

(a) $\phi > 0$ (b) $\phi < 0$

Figure 2.3   *Autocorrelation Functions for AR(1) Processes*

successive observations tend to be relatively small. A negative ϕ leads to a very irregular pattern, since adjacent observations are negatively correlated.

The variance and autocovariance of the AR(2) process may be obtained by generalising (2.28). Multiplying (2.8) by $y_{t-\tau}$ and taking expectations yields

$$E(y_t y_{t-\tau}) = \phi_1 E(y_{t-1} y_{t-\tau}) + \phi_2 E(y_{t-2} y_{t-\tau}) + E(\epsilon_t y_{t-\tau}).$$
(2.31)

The last term is zero for $\tau > 0$ and so

$$\gamma(\tau) = \phi_1 \gamma(\tau - 1) + \phi_2 \gamma(\tau - 2), \qquad \tau = 1, 2, \ldots \qquad (2.32)$$

Dividing through by $\gamma(0)$ gives a second-order difference equation for the autocorrelation function, i.e.

$$\rho(\tau) = \phi_1 \rho(\tau - 1) + \phi_2 \rho(\tau - 2), \qquad \tau = 1, 2, \ldots \qquad (2.33)$$

Since $\rho(-1) = \rho(1)$, setting $\tau = 1$ yields

$$\rho(1) = \phi_1 + \phi_2 \rho(1)$$

and so starting values for the autocorrelation function are given by $\rho(0) = 1$ and $\rho(1) = \phi_1/(1 - \phi_2)$. Equation (2.33) is of exactly the same form as the homogeneous equation (2.9). Its solution will therefore exhibit the same characteristics as the deterministic component of the AR(2) process. In particular, for complex roots it may show damped cyclical behaviour as illustrated in figure 2.4.

The variance of the AR(2) process may be obtained from (2.31) with $\tau = 0$. The last term is no longer zero as

$$E(\epsilon_t y_t) = E[\epsilon_t(\phi_1 y_{t-1} + \phi_2 y_{t-2} + \epsilon_t)]$$
$$= \phi_1 E(\epsilon_t y_{t-1}) + \phi_2 E(\epsilon_t y_{t-2}) + E(\epsilon_t^2)$$
$$= 0 + 0 + \sigma^2 = \sigma^2.$$

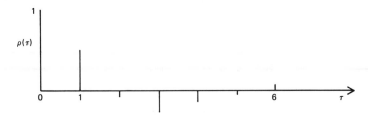

Figure 2.4 *Autocorrelation Function for an AR(2) Process with Complex Roots*

and so

$$\gamma(0) = \phi_1 \gamma(1) + \phi_2 \gamma(2) + \sigma^2. \tag{2.34}$$

Dividing through by $\gamma(0)$ and rearranging gives

$$\gamma(0) = \frac{\sigma^2}{1 - \rho(1)\phi_1 - \rho(2)\phi_2}, \tag{2.35}$$

but since $\rho(1) = \phi_1/(1 - \phi_2)$ and $\rho(2) = \phi_1 \rho(1) + \phi_2$, $\gamma(0)$ may be expressed in terms of ϕ_1 and ϕ_2 as

$$\gamma(0) = \left(\frac{1 - \phi_2}{1 + \phi_2}\right) \frac{\sigma^2}{[(1 - \phi_2)^2 - \phi_1^2]}. \tag{2.36}$$

The same techniques may be used to derive the time domain properties of any AR process. Multiplying (2.1) by $y_{t-\tau}$, taking expectations and dividing by $\gamma(0)$ gives the pth order difference equation

$$\rho(\tau) = \phi_1 \rho(\tau - 1) + \cdots + \phi_p \rho(\tau - p), \qquad \tau = 1, 2, \ldots \tag{2.37}$$

When $\tau = 0$, the following expression for the variance is obtained after some rearrangement:

$$\gamma(0) = \sigma^2/[1 - \rho(1)\phi_1 - \cdots - \rho(p)\phi_p]. \tag{2.38}$$

For all stationary AR(p) processes the autocorrelation function 'damps down', in the sense that $\rho(\tau)$ tends to zero as $\tau \to \infty$. The actual behaviour it exhibits, for example with regard to cyclical movements, depends on the roots of the characteristic equation, (2.25).

3. Moving Average Processes

A moving average process of order q is written as

$$y_t = \epsilon_t + \theta_1 \epsilon_{t-1} + \cdots + \theta_q \epsilon_{t-q}, \qquad t = 1, \ldots, T, \tag{3.1}$$

and denoted by $y_t \sim MA(q)$. A finite moving average process is always stationary. Condition (1.24) is satisfied automatically and the covariances can be derived without placing any restrictions on the parameters $\theta_1, \ldots, \theta_q$. However, it is sometimes necessary to express an MA process in autoregressive form. If this is to be done, the MA parameters must satisfy conditions similar to those which have to be imposed on AR parameters in order to ensure stationarity. If these conditions are satisfied, the MA process is said to be invertible.

Autocovariance and Autocorrelation Functions

Taking expectations in (3.1) immediately shows that y_t has zero mean, while

$$\gamma(0) = E(y_t^2) = (1 + \theta_1^2 + \cdots + \theta_q^2)\sigma^2. \tag{3.2}$$

The autocovariances are given by

$$\gamma(\tau) = \begin{cases} \theta_\tau + \theta_1\theta_{\tau+1} + \cdots + \theta_{q-\tau}\theta_q)\sigma^2, & \tau = 1, \ldots, q, \\ 0 & , & \tau > q; \end{cases} \tag{3.3}$$

cf. (1.14c) and (1.14d). Since the autocovariances at lags greater than q are zero, the autocovariance function, and therefore the autocorrelation function, has a distinct 'cut-off' at $\tau = q$. This contrasts with the autocovariance function for an AR process which gradually decays towards zero. If q is small, the maximum value of $|\rho(1)|$ is well below unity. It can be shown that

$$|\rho(1)| \leqslant \cos[\pi/(q + 2)]. \tag{3.4}$$

For $q = 1$, this means that the maximum value of $|\rho(1)|$ is 0.5, a point which can be verified directly from (1.15).

Invertibility

The MA(1) process, (1.1), can be expressed in terms of lagged values of y_t by substituting repeatedly for lagged values of ϵ_t. This yields

$$y_t = \theta y_{t-1} - \theta^2 y_{t-2} + \cdots - (-\theta)^J y_{t-J} + \epsilon_t - (-\theta)^{J+1}\epsilon_{t-J-1}; \tag{3.5}$$

cf. (2.3). If y_t is not to depend on a shock to the system arising at some point in the remote past, θ must be less than one in absolute value. If J is allowed to go to infinity, the last term in (3.5) disappears and y_t may be written as an infinite autoregressive process with declining weights, i.e.

$$y_t = - \sum_{j=0}^{\infty} (-\theta)^j y_{t-j} + \epsilon_t. \tag{3.6}$$

An MA(1) model with $|\theta| > 1$ is not invertible, but it is still stationary. However, its autocorrelation function can be reproduced exactly by an invertible process with parameter $1/\theta$. This can be seen by substituting in (1.15) to give

$$\rho(1) = \frac{1/\theta}{1 + (1/\theta)^2} = \frac{\theta}{1 + \theta^2}. \tag{3.7}$$

Except for the case of $|\theta| = 1$, therefore, a particular autocorrelation function will be compatible with two processes, only one of which is invertible. Restricting attention to invertible processes resolves the problem of identifiability, but the fundamental reason for dismissing non-invertible processes is that they give rise to inefficient predictions; see Section 6.4.

An MA(1) process with $|\theta| = 1$ is something of an anomaly, since it may be uniquely identified from the autocorrelation function. Although such processes are not *strictly* invertible, they cannot be dismissed along with processes for which $|\theta| > 1$, since perfectly sensible predictions can be based on them.

The concept of invertibility extends to higher order MA processes. The conditions necessary for invertibility may be expressed in terms of the associated MA polynomial, (1.26b), by requiring that the roots of

$$\theta(L) = 0 \tag{3.8}$$

lie outside the unit circle.

4. Mixed Processes

The general autoregressive-moving average process was defined in (1.25) and such a process is indicated by the notation $y_t \sim$ ARMA(p, q). The AR(p) and MA(q) models are special cases and it is quite legitimate to denote them by ARMA$(p, 0)$ and ARMA$(0, q)$ respectively. However, a mixed process will always be taken to mean one in which both p and q are greater than zero.

Stationarity and Invertibility

Whether or not a mixed process is stationary depends solely on its autoregressive part. The conditions may be expressed in terms of the autoregressive polynomial associated with (1.27) by requiring that the roots of $\phi(L) = 0$ lie outside the unit circle. In a similar way, the invertibility condition is exactly the same as for an MA(q) process, namely that the roots of (3.8) should lie outside the unit circle.

The reason why stationarity depends on the AR part of the model becomes apparent as soon as an attempt is made to express a mixed process as an infinite moving average. The ARMA(1, 1) process

$$y_t = \phi y_{t-1} + \epsilon_t + \theta \epsilon_{t-1}, \qquad t = 1, \ldots, T, \tag{4.1}$$

is the simplest mixed process and it would be possible to substitute repeatedly for lagged values of y_t as in (2.3). An alternative way of going about this is to write (4.1) in the form

$$(1 - \phi L)y_t = (1 + \theta L)\epsilon_t \tag{4.2}$$

and to divide both sides by $(1 - \phi L)$. This yields

$$y_t = \frac{\epsilon_t}{1 - \phi L} + \frac{\theta \epsilon_{t-1}}{1 - \phi L}. \tag{4.3}$$

If L is deemed to satisfy the condition $|L| \leqslant 1$, the term $1/(1 - \phi L)$ can be regarded as the sum of the infinite geometric progression $1, \phi L, (\phi L)^2, \ldots$ when $|\phi| < 1$, and so (4.3) can be re-written as

$$y_t = \sum_{j=0}^{\infty} (\phi L)^j \epsilon_t + \theta \sum_{j=0}^{\infty} (\phi L)^j \epsilon_{t-1}$$

$$= \sum_{j=0}^{\infty} \phi^j \epsilon_{t-j} + \theta \sum_{j=0}^{\infty} \phi^j \epsilon_{t-j-1}$$

$$= \epsilon_t + \sum_{j=1}^{\infty} (\theta \phi^{j-1} + \phi^j)\epsilon_{t-j}. \tag{4.4}$$

When $\theta = 0$, this expression reduces to (2.5). Given that $|\phi| < 1$, the weights in (4.4) decline sufficiently rapidly for the process to have finite variance and for the autocovariances to exist.

For $p > 1$, the infinite MA representation may be obtained by a similar device which entails factorising $\phi(L)$ and expanding $\phi^{-1}(L)\theta(L)$ in partial fractions. A more convenient way of proceeding is to equate coefficients of powers of L in the expression

$$\theta(L) = \phi(L)\psi(L), \tag{4.5}$$

where $\theta(L)$ and $\phi(L)$ are the polynomials defined in (1.26). The infinite MA polynomial, $\psi(L)$, was defined in (2.17) and expression (4.5) is simply a generalisation of the equation (2.20) used to obtain MA coefficients in the AR(2) case. After some re-arrangement (4.5) yields

$$\psi_0 = 1$$

$$\psi_j = \theta_j + \sum_{i=1}^{\min(j, p)} \phi_i \psi_{j-i}, \qquad j = 1, \ldots, q$$

$$\psi_j = \sum_{i=1}^{\min(j, p)} \phi_i \psi_{j-i}, \qquad j > q. \tag{4.6}$$

For $j \geqslant \max(p, q + 1)$, the ψ_j's are determined by the difference equation, (4.6), with starting values given by the previous p values of ψ_j.

Example 1   In the AR(2) model it was shown earlier that the MA coefficients depend on the difference equation (2.22) for $j \geqslant 2$, with starting values $\psi_0 = 1$ and $\psi_1 = \phi_1$.

Example 2   For the ARMA(1, 1) model, (4.1), the ψ_j's are computed from the difference equation

$$\psi_j = \phi \psi_{j-1}, \qquad j \geqslant 2, \tag{4.7}$$

with starting value

$$\psi_1 = \theta + \phi \psi_0 = \theta + \phi. \tag{4.8}$$

Similar techniques may be used to obtain the infinite autoregressive representation of an invertible ARMA process.

Autovariance and Autocorrelation Functions

The time domain properties of mixed models incorporate features of both AR and MA processes. This is illustrated clearly by the ARMA(1, 1) model. Multiplying (4.1) through by $y_{t-\tau}$ and taking expectations gives

$$\gamma(\tau) = \phi \gamma(\tau - 1) + E(\epsilon_t y_{t-\tau}) + \theta E(\epsilon_{t-1} y_{t-\tau}),$$

$$\tau = 0, 1, 2, \ldots \quad (4.9)$$

The last two expectations are zero for $\tau > 1$. For $\tau = 1$, the second of the expectations becomes

$$E(\epsilon_{t-1} y_{t-1}) = E[\epsilon_{t-1}(\phi y_{t-2} + \epsilon_{t-1} + \theta \epsilon_{t-2})] = \sigma^2,$$

although the first remains zero. When $\tau = 0$, both expectations are non-zero, being given by

$$E(\epsilon_t y_t) = \sigma^2,$$

and

$$\begin{aligned}
E(\epsilon_{t-1} y_t) &= E[\epsilon_{t-1}(\phi y_{t-1} + \epsilon_t + \theta \epsilon_{t-1})] \\
&= \phi E(\epsilon_{t-1} y_{t-1}) + \theta \sigma^2 \\
&= \phi \sigma^2 + \theta \sigma^2.
\end{aligned}$$

The autocovariance function is therefore

$$\gamma(0) = \phi \gamma(1) + \sigma^2 + \theta \phi \sigma^2 + \theta^2 \sigma^2 \tag{4.10a}$$

$$\gamma(1) = \phi \gamma(0) + \theta \sigma^2 \tag{4.10b}$$

$$\gamma(\tau) = \phi\gamma(\tau - 1), \qquad \tau = 2, 3, \ldots \tag{4.10c}$$

Substituting for $\gamma(1)$ from the second of these equations into the first yields

$$\gamma(0) = \frac{1 + \theta^2 + 2\phi\theta}{1 - \phi^2} \sigma^2, \tag{4.11}$$

and so

$$\gamma(1) = \frac{(1 + \phi\theta)(\phi + \theta)}{1 - \phi^2} \sigma^2. \tag{4.12}$$

Dividing (4.12) and (4.10c) by (4.11) gives the autocorrelation function

$$\rho(1) = \frac{(1 + \phi\theta)(\phi + \theta)}{1 + \theta^2 + 2\phi\theta} \tag{4.13a}$$

$$\rho(\tau) = \phi\rho(\tau - 1), \qquad \tau = 2, 3, \ldots \tag{4.13b}$$

On inspecting the autocorrelation function, it will be seen that, for $\tau > 1$, its behaviour is governed by the first-order difference equation (4.13b). Thus the autocorrelations exhibit exponential decay, with oscillations when ϕ is negative. This is exactly as in the AR(1) case. However, while the starting value for the difference equation in the AR(1) model is simply $\rho(0) = 1$, in the ARMA(1, 1) model the starting value is given by $\rho(1)$. As expression (4.13a) makes clear, $\rho(1)$ depends on both ϕ and θ, and the sign of $\rho(1)$ will depend on the sign of $(\phi + \theta)$.

As an illustration of the kind of pattern exhibited by the auto-correlation function, suppose $\phi = 0.3$ and $\theta = 0.9$. In this case $\rho(1) = 0.649$, and the autocorrelations follow the scheme in figure 2.5(a). Changing the moving average parameter to $\theta = -0.9$ gives $\rho(1) = -0.345$. The autocorrelations once again decay exponentially, but from a negative rather than a positive starting value. This is shown in figure 2.5(b). All in all there are six distinct patterns for the autocorrelation function from an ARMA(1, 1) model, depending on the different values taken by ϕ and θ.

The properties of higher order ARMA models may be derived in a similar fashion. The general pattern which emerges for the auto-correlation function is that the first q autocorrelations depend on both the moving average and the autoregressive parameters. Higher order autocorrelations are given by a pth order difference equation of the form (2.37), with $\rho(q), \rho(q - 1), \ldots, \rho(q - p + 1)$ as starting values.

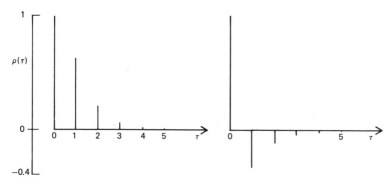

(a) $\phi = 0.3, \theta = 0.9$ (b) $\phi = 0.3, \theta = -0.9$

Figure 2.5   *Autocorrelation Functions for ARMA(1, 1) Processes*

Common Factors

If the AR and MA polynomials in (1.27) have a root which is the same, they are said to have a *common factor*. In this case the model is over-parameterised, since a model with identical properties can be constructed by reducing both p and q by one. It is important to recognise common factors, since if they exist the model will not be identifiable and computational problems may arise; see EATS (Section 3.6).

Example 3   In the ARMA(2, 1) model

$$y_t = 0.2y_{t-1} + 0.15y_{t-2} + \epsilon_t + 0.3\epsilon_{t-1}, \tag{4.14}$$

the AR polynomial may be factorised as

$$(1 - 0.2L - 0.15L^2) = (1 - 0.5L)(1 + 0.3L).$$

On re-writing the model in the form

$$y_t = \phi^{-1}(L)\theta(L)\epsilon_t = \frac{(1 + 0.3L)}{(1 - 0.5L)(1 + 0.3L)}\,\epsilon_t$$

it is immediately apparent that (4.14) has exactly the same MA representation as the AR(1) model

$$y_t = 0.5y_{t-1} + \epsilon_t. \tag{4.15}$$

Hence (4.15) and (4.14) have identical autocovariance functions and so (4.14) is over-parameterised.

5. Aggregation and Missing Observations*

There are two types of aggregation. The first arises when the observed series is the sum of two or more individual series. These individual series are known as *unobserved components* and if they follow stationary ARMA processes a natural question to ask is whether the aggregated series follows an ARMA process as well.

The second type of aggregation is aggregation over time. The available data may refer to a time unit greater than the time unit employed in the model. Thus the 'natural' time interval in the model specification may be one quarter, whereas data is only available on an annual basis. If the variable in question is a flow, such as Gross National Product, then it may be regarded as having been subjected to *temporal aggregation*. If the disaggregated observations are generated by an ARMA process, the relevant question is then whether the aggregated observations are also generated by an ARMA process and, if so, what are its properties. When the variable in question is a stock, such as the money supply, the nature of the problem is slightly different. A stock variable is essentially continuous, but it may be sensible to model it in terms of a particular unit of time. If the measurements on the variable are made at less frequent intervals than the model requires, the problem is one of *missing observations*. There is no aggregate information available, although the nature of the problem is not dissimilar to the one which arises with flow variables. In fact temporal aggregation is often treated under the heading of missing observations.

The sub-sections below set out certain results on the properties of the observed series when aggregation has taken place or there are missing observations. Some of the methods which can be used for extracting individual components or estimating disaggregated observations are touched on very briefly in Chapter 4.

Aggregation of Different Series

Consider two series, y_t and x_t, which are stationary, independent and have zero mean. If y_t is ARMA(p_1, q_1) and x_t is ARMA(p_2, q_2) the sum

$$z_t = y_t + x_t, \tag{5.1}$$

is ARMA(p_3, q_3), where $p_3 \leqslant p_1 + p_2$ and $q_3 \leqslant \max(p_1 + q_2, p_2 + q_1)$. A proof will be found in Granger and Morris (1976).

The inequalities in the expressions for q_3 and p_3 are necessary because some simplification may occur in special cases.

Example 1   Suppose that x_t and y_t are independent AR(1) processes with a common AR parameter, i.e.

$$y_t = \phi y_{t-1} + \epsilon_t, \tag{5.2a}$$

$$x_t = \phi x_{t-1} + \eta_t, \tag{5.2b}$$

where ϵ_t and η_t are independent white noise processes. Then

$$y_t + x_t = \phi(y_{t-1} + x_{t-1}) + \epsilon_t + \eta_t, \tag{5.3}$$

and so

$$z_t = \phi z_{t-1} + \xi_t, \tag{5.4}$$

where $\xi_t = \epsilon_t + \eta_t$ is white noise. Thus, z_t is AR(1) rather than ARMA(2, 1).

It is worth setting out a number of specific cases of the general result. If there is no coincidental reduction in parameters, as in the example above, these cases may be enumerated as follows:

$$AR(p) + \text{white noise} = ARMA(p, p) \tag{5.5a}$$

$$AR(p_1) + AR(p_2) = ARMA[p_1 + p_2, \max(p_1, p_2)] \tag{5.5b}$$

$$MA(q_1) + MA(q_2) = ARMA[0, \max(q_1, q_2)] \tag{5.5c}$$

$$ARMA(p, q) + \text{white noise} = ARMA[p, \max(p, q)] \tag{5.5d}$$

$$AR(p) + MA(q) = ARMA(p, p + q) \tag{5.5e}$$

It is interesting to note that the addition of white noise to an AR process leads to a mixed process, while adding white noise to an MA process yields an MA process of exactly the same order. It is relatively easy to demonstrate this last result directly.

Example 2   If an independent white noise series, η_t, is added to the MA(1) process, (1.1), the result is

$$z_t = \epsilon_t + \theta \epsilon_{t-1} + \eta_t. \tag{5.6}$$

If $\text{Var}(\epsilon_t) = \sigma_1^2$ while $\text{Var}(\eta_t) = \sigma_2^2$, the autocorrelation function of z_t is

$$\rho(\tau) = \begin{cases} \sigma_1^2 \theta / [\sigma_1^2(1 + \theta^2) + \sigma_2^2], & \tau = 1 \\ 0 & , & \tau \geq 2. \end{cases} \tag{5.7}$$

This has exactly the same form as the autocorrelation function of an MA(1) process. Thus z_t is an MA(1) process. Its parameter can be found by solving (1.15).

Temporal Aggregation and Missing Observations

Suppose that y_t is a flow variable which obeys an ARMA(p, q) process, but that it is observed in aggregate form every m time period. The observations are therefore defined by

$$y_t^* = \sum_{j=0}^{m-1} y_{t-j}, \qquad t = m, 2m, 3m, \ldots \qquad (5.8)$$

and it can be shown that they follow an ARMA(p, q^*) process where q^* is given by the expression

$$[(p + 1)(m - 1) + q]/m$$

rounded down to the nearest integer; see Brewer (1973) and Amemiya and Wu (1972).

Example 3   If y_t is an AR(1) process whose aggregate value is observed every other time period, then

$$y_t = \phi(\phi y_{t-2} + \epsilon_{t-1}) + \epsilon_t$$
$$= \phi^2 y_{t-2} + \phi \epsilon_{t-1} + \epsilon_t, \qquad (5.9)$$

while

$$y_{t-1} = \phi^2 y_{t-3} + \phi \epsilon_{t-2} + \epsilon_{t-1}. \qquad (5.10)$$

Summing (5.9) and (5.10) yields an expression for the process generating the aggregate series, i.e.

$$y_t + y_{t-1} = \phi^2(y_{t-2} + y_{t-3}) + \epsilon_t + (1 + \phi)\epsilon_{t-1} + \phi \epsilon_{t-2},$$
$$t = 2, 4, 6, \ldots \quad (5.11)$$

This can be re-written as

$$y_\tau^* = \phi^2 y_{\tau-1}^* + u_\tau, \qquad \tau = 2t, \quad \tau = 1, 2, 3, \ldots \qquad (5.12)$$

where u_τ has an autocorrelation function appropriate to an MA(1) process. Thus y_τ^* is ARMA(1, 1).

As m increases, the aggregated series tends towards ARMA($p, p + 1$) if $q \geqslant p + 1$ and towards ARMA(p, p) if $q < p + 1$. However, the autocorrelations between successive observations also tend to become weaker and so the aggregate series moves closer to white noise.

Similar results have been derived by Brewer (1973) for a stock variable. If the process is ARMA(p, q) with respect to a particular time period, but is observed only every mth time period, the observations follow an ARMA(p, q^*) process, where q^* is given by the expression

$$[p(m-1)+q]/m$$

rounded down to the nearest integer. As the sampling interval, m, becomes large, the process followed by the observations converges to ARMA(p, p) if $q \geqslant p$ and to ARMA$(p, p-1)$ if $q < p$.

Example 4   If $y_t \sim$ AR(1), but is only observed every other time period, the observations also follow an AR(1) process. This can be seen immediately on substituting for y_{t-1}, i.e.

$$y_t = \phi^2 y_{t-2} + \{\epsilon_t + \phi\epsilon_{t-1}\}. \tag{5.13}$$

The term in curly brackets is white noise with respect to the time periods $t = 2, 4, 6, \ldots$

6. Multivariate Time Series

The concepts involved in analysing univariate time series may be extended to multivariate series. Attention is focused on the joint behaviour of the elements in the $N \times 1$ vector $y_t = (y_{1t}, \ldots, y_{Nt})'$, observed over the period $t = 1, \ldots, T$. Having analysed this behaviour, an attempt may be made to model y_t by a multivariate, or vector, ARMA process.

Stationarity

It may be assumed, without any loss of generality, that $E(y_t) = 0$. The covariance matrix between y_t and $y_{t-\tau}$ is then defined by $E(y_t y'_{t-\tau})$. The ijth element in this matrix is equal to $E(y_{it} y_{j, t-\tau})$ and the vector process y_t is covariance stationary only if all such elements are independent of t for all values of τ. This is a stronger condition than simply requiring that each individual series be covariance stationary. This would merely constrain the diagonal elements in the covariance matrices to be independent of t, without imposing a similar constraint on the off-diagonal elements. These off-diagonal elements denote the *cross covariances* between different series at particular lags, and two stationary series will only be *jointly stationary* if these quantities are independent of t.

For a weakly stationary vector process, the covariance between y_t and $y_{t-\tau}$ will be denoted by

$$E(y_t y'_{t-\tau}) = \Gamma(\tau), \qquad \tau = 0, 1, 2, \ldots \tag{6.1}$$

and $\Gamma(\tau)$ will generally be referred to as the *autocovariance matrix* at lag τ. The *covariance* matrix is obtained when $\tau = 0$; this matrix is symmetric.

In the univariate case the autocovariance function has the property that $\gamma(-\tau) = \gamma(\tau)$, but for multivariate models the matrices $\Gamma(\tau)$ and $\Gamma(-\tau)$ will not, in general, be identical. However, taking the ijth element in $\Gamma(\tau)$ and setting $t = t^* + \tau$ gives

$$\gamma_{ij}(\tau) = E(y_{it} y_{j, t-\tau}) = E(y_{i, t^*+\tau} y_{j, t^*}) = E(y_{jt^*} y_{i, t^*+\tau}) = \gamma_{ji}(-\tau) \tag{6.2}$$

and so the ijth element in $\Gamma(\tau)$ is equal to the jith element in $\Gamma(-\tau)$. Thus the autocovariance matrix of a stationary vector process satisfies the relation

$$\Gamma(\tau) = \Gamma'(-\tau), \qquad \tau = 0, 1, 2, \ldots \tag{6.3}$$

An autocovariance matrix may be standardised to yield the corresponding *autocorrelation matrix*. The ijth element in $\Gamma(\tau)$ is simply divided by the square roots of the variances of y_i and y_j to give the corresponding autocorrelation or cross-correlation. In matrix terms the autocorrelation matrix at lag τ, $P(\tau)$, is defined by

$$P(\tau) = D_0^{-1} \Gamma(\tau) D_0^{-1}, \qquad \tau = 0, \pm 1, \pm 2, \ldots, \tag{6.4}$$

where $D_0^2 = \text{diag}\{\gamma_{11}(0), \ldots, \gamma_{NN}(0)\}$. The ijth element of $P(\tau)$ gives the cross-correlation between y_i and y_j at lag τ, i.e.

$$\rho_{ij}(\tau) = \gamma_{ij}(\tau)/\sqrt{\gamma_{ii}(0)\gamma_{jj}(0)}. \tag{6.5}$$

Cross-Correlation Function

The information on the relationship between different variables in y_t is contained in the off-diagonal elements of the autocorrelation matrices. For a *bivariate* series, the structure of these matrices means that there is only a single quantity to consider, and a plot of this quantity against both negative and positive values of τ gives the cross-correlation function.

Example 1 Suppose that ϵ_t and x_t are independent white noise processes with mean zero and variances σ^2 and σ_x^2 respectively. A third variable, y_t, is defined by the equation

$$y_t = \beta x_{t-1} + \epsilon_t. \tag{6.6}$$

The vector process $(y_t, x_t)'$ is covariance stationary with the cross-correlation function given by

$$\rho_{yx}(\tau) = \begin{cases} \beta\sigma_x/(\beta^2\sigma_x^2 + \sigma^2)^{1/2}, & \tau = 1 \\ 0, & \tau \neq 1 \end{cases} \tag{6.7}$$

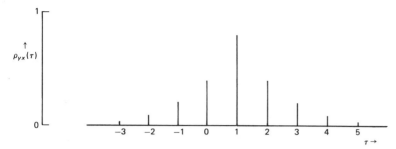

Figure 2.6   *Cross-Correlation Function defined in (2.6.10)*

Example 2   If, in the above example, x_t is generated by a stationary AR(1) process,

$$x_t = \phi x_{t-1} + \eta_t, \qquad |\phi| < 1, \tag{6.8}$$

where η_t is white noise with mean zero and variance, σ_η^2, the cross-correlation function is more complicated. Given that ϵ_t and η_t are independent, multiplying (6.6) by $x_{t-\tau}$ and taking expectations yields

$$\gamma_{yx}(\tau) = \beta \gamma_x(\tau - 1), \qquad \tau = 0, \pm 1, \pm 2, \ldots \tag{6.9}$$

Thus

$$\rho_{yx}(\tau) = \beta \phi^{|\tau - 1|} \sigma_x / \sigma_y, \qquad \tau = 0, \pm 1, \ldots \tag{6.10}$$

where $\sigma_x^2 = \sigma_\eta^2 / (1 - \phi^2)$ and $\sigma_y^2 = \beta^2 \sigma_x^2 + \sigma^2$; see figure 2.6. Although the relationship between x_t and y_t shows up quite clearly in (6.7), this is no longer the case when x_t is serially correlated. As (6.10) shows, the cross-correlation function depends not only on the relationship between x_t and y_t, but also on the nature of the process generating x_t.

7. Vector Autoregressive-Moving Average Processes

An indeterministic vector process, y_t, can always be written as a vector MA process of infinite order, i.e.

$$y_t = \sum_{j=0}^{\infty} \Psi_j \epsilon_{t-j} \tag{7.1}$$

where Ψ_j is an $N \times N$ matrix of parameters, $\psi_0 = I$, and ϵ_t is an $N \times 1$ vector of white noise variables with mean zero. The properties of ϵ_t, which is sometimes known as a *multivariate white noise process*, may be summarised as follows:

$$E(\epsilon_t) = 0$$

and

$$E(\epsilon_t \epsilon_s') = \begin{cases} \Omega, & t = s \\ 0, & t \neq s \end{cases} \qquad (7.2)$$

where Ω is an $N \times N$ covariance matrix. Thus the elements of ϵ_t may be contemporaneously correlated.

The moving average representation in (7.1) is a direct generalisation of (1.23). As in the univariate case, y_t may be modelled parsimoniously by a mixed process of the form

$$y_t = \Phi_1 y_{t-1} + \cdots + \Phi_p y_{t-p} + \epsilon_t + \cdots + \Theta_q \epsilon_{t-q} \qquad (7.3)$$

where the Φ_i's are $N \times N$ matrices of AR parameters and the Θ_j's are $N \times N$ matrices of MA parameters. This model is known as a multivariate, or vector, ARMA(p, q) process.

Autocovariance and Autocorrelation Functions

The time domain properties of vector ARMA processes may be derived using similar techniques to those employed for univariate processes.

Example 1   The vector MA(1) process,

$$y_t = \epsilon_t + \Theta \epsilon_{t-1}, \qquad (7.4)$$

is stationary, irrespective of the values of the parameters in Θ. Since $E(y_t) = 0$, the covariance matrix of y_t is

$$\Gamma(0) = E(y_t y_t') = \Omega + \Theta \Omega \Theta', \qquad (7.5)$$

while the autocovariance matrices are given by

$$\Gamma(\tau) = E(y_t y_{t-\tau}') = \begin{cases} \Theta \Omega, & \tau = 1 \\ \Omega \Theta', & \tau = -1 \\ 0, & |\tau| \geqslant 2. \end{cases} \qquad (7.6)$$

Note that $\Gamma'(-1) = \Gamma(1)$, but that there is an abrupt cut-off for $|\tau| \geqslant 2$. This behaviour exactly parallels that in the univariate MA(1) process.

Example 2   The conditions under which the first-order vector autoregressive process,

$$y_t = \Phi y_{t-1} + \epsilon_t, \qquad (7.7)$$

is stationary are derived in the next sub-section. Given that these conditions are satisfied, the autocovariance matrices may be derived by post-multiplying y_t by $y_{t-\tau}$ and taking expectations. Thus

$$E(y_t y'_{t-\tau}) = \Phi E(y_{t-1} y'_{t-\tau}) + E(\epsilon_t y_{t-\tau}). \tag{7.8}$$

For $\tau \geqslant 1$, the last term is a vector of zeros, and so

$$\Gamma(\tau) = \Phi \Gamma(\tau - 1), \qquad \tau \geqslant 1. \tag{7.9}$$

This is a vector difference equation with solution

$$\Gamma(\tau) = \Phi^\tau \Gamma(0), \qquad \tau \geqslant 0; \tag{7.10}$$

cf. (2.29).

When $\tau = 0$, the last term in (7.8) is given by

$$E[\epsilon_t(\Phi y_{t-1} + \epsilon_t)'] = E(\epsilon_t \epsilon'_t) = \Omega.$$

Furthermore, in view of (6.3) and (7.10), the first term on the right-hand side of (7.8) becomes

$$\Phi \Gamma(-1) = \Phi \Gamma'(1) = \Phi \Gamma(0)\Phi',$$

and so

$$\Gamma(0) = \Phi \Gamma(0)\Phi' + \Omega. \tag{7.11}$$

There are a number of ways of solving these equations in order to evaluate $\Gamma(0)$. One possibility is to adopt an iterative procedure in which successive approximations are computed using an initial guess for $\Gamma(0)$. Alternatively, a direct solution can be obtained from the identity

$$\text{vec}\{\Gamma(0)\} = [I - \Phi \otimes \Phi]^{-1} \text{vec}\{\Omega\}. \tag{7.12}$$

The symbol $\otimes$ denotes a Kronecker product while the vec(.) operator indicates that the columns of an $n \times m$ matrix are being stacked, one on top of the other, to form a vector of length nm.

Finally it is worth noting that in the model defined by (6.6) and (6.8), $(y_t\ x_t)'$ is a vector AR(1) process with

$$\Phi = \begin{bmatrix} 0 & \beta \\ 0 & \phi \end{bmatrix}.$$

Although the variances and cross-covariances can be derived directly in this case, obtaining them from the general formulae above provides a useful exercise.

Stationarity and Invertibility for Multivariate Autoregressive-Moving Average Processes*

Repeatedly substituting for lagged values of y_t in the vector AR(1) process, (7.7) gives

$$y_t = \sum_{j=0}^{J-1} \Phi^j \epsilon_{t-j} + \Phi^J y_{t-J}. \tag{7.13}$$

By analogy with the argument set out for the univariate case, y_t ceases to depend on y_{t-J} if the elements of Φ^J tend to zero as J becomes large. In order to determine the conditions under which this holds, we will attempt to diagonalise the matrix Φ, by trying to find a nonsingular $N \times N$ matrix, Q, such that

$$\Phi = Q\Lambda Q^{-1}, \tag{7.14}$$

where $\Lambda = \text{diag}\{\lambda_1, \ldots, \lambda_N\}$. The elements $\lambda_1, \ldots, \lambda_N$, are obtained by solving the *determinantal equation*

$$|\Phi - \lambda I| = 0. \tag{7.15}$$

Thus if $N = 2$ and ϕ_{ij} is the ijth element in Φ, the determinantal equation is

$$\begin{vmatrix} \phi_{11} - \lambda & \phi_{12} \\ \phi_{21} & \phi_{22} - \lambda \end{vmatrix} = 0 \tag{7.16}$$

and the two roots are obtained by solving the quadratic

$$\lambda^2 - (\phi_{11} + \phi_{22})\lambda + (\phi_{11}\phi_{22} - \phi_{12}\phi_{21}) = 0. \tag{7.17}$$

The roots may be real or a pair of complex conjugates.

The diagonalisation implied by (7.14) is only possible if all N roots are distinct. Assuming this to be the case,

$$\Phi^J = Q\Lambda Q^{-1} \cdot Q\Lambda Q^{-1} \ldots Q\Lambda Q^{-1} = Q\Lambda^J Q^{-1}.$$

Since Q is independent of J while $\Delta^J = \text{diag}(\lambda_1^J, \ldots, \lambda_N^J)$, it follows that

$$\lim_{J \to \infty} \Phi^J = \lim_{J \to \infty} Q\Lambda^J Q^{-1} = 0, \tag{7.18}$$

provided that the roots of the determinantal equation are less than one in absolute value. Thus the condition

$$|\lambda_i| < 1, \qquad i = 1, \ldots, N, \tag{7.19}$$

is necessary for (7.7) to be stationary.

This result generalises to a process with a pth order autoregressive component. The model can be shown to be stationary if the roots of

$$|\lambda^p I - \lambda^{p-1}\Phi_1 - \cdots - \Phi_p| = 0, \tag{7.20a}$$

are less than one in absolute value. Similarly, it is invertible if the roots of

$$|\lambda^q I + \lambda^{q-1}\Theta_1 + \cdots + \Theta_q| = 0 \tag{7.20b}$$

are less than one in absolute value.

The lag operator representation of (7.3) is

$$\Phi_p(L)y_t = \Theta_q(L)\epsilon_t, \tag{7.21}$$

where

$$\Phi_p(L) = I - \Phi_1 L - \cdots - \Phi_p L^p \tag{7.22a}$$

and

$$\Theta_q(L) = I + \Theta_1 L + \cdots + \Theta_q L^q. \tag{7.22b}$$

The stationarity and invertibility conditions may be expressed in terms of the *determinantal polynomials*, $|\Phi_p(L)|$ and $|\Theta_q(L)|$, although the conditions must now be stated in terms of roots lying *outside* the unit circle, just as in the univariate case.

Identifiability*

In Section 4 attention was drawn to the implications of common factors for the identifiability of univariate ARMA models. Similar considerations arise in multivariate models, although the conditions for identifiability are much more complex. Reference should be made to Hannan (1969) or Granger and Newbold (1977, pp. 219–224).

Multivariate and Univariate Models

Any stationary vector ARMA process may be written in the form (7.1) by inverting the matrix polynomial $\Phi(L)$. This yields

$$y_t = \Phi^{-1}(L)\Theta(L)\epsilon_t. \tag{7.23}$$

The power series expansion of $\Phi^{-1}(L)$ is

$$\Phi^{-1}(L) = \Phi^\dagger(L)/|\Phi(L)|, \tag{7.24}$$

where $\Phi^\dagger(L)$ is the adjoint matrix of $\Phi(L)$. Substituting into (7.23) and re-arranging gives the *autoregressive final form*,

$$|\Phi(L)|y_t = \Phi^\dagger(L)\Theta(L)\epsilon_t. \tag{7.25}$$

The right-hand side of each of the N equations in (7.25) consists of a linear combination of MA processes each of order $p + q$. However, each of these linear combinations can be expressed as an MA process of order $p + q$; cf. Example 5.2. Thus the equation in (7.25) can be written as

$$\phi_i^*(L)y_t = \theta_i^*(L)\eta_{it}, \qquad i = 1, \ldots, N, \tag{7.26}$$

where η_{it} is white noise, $\theta_i^*(L)$ is an MA polynomial in the lag operator of order $p + q$ and $\phi_i^*(L) = |\Phi(L)|$ is an AR polynomial of order $2p$.

The implication of this result is that each variable in y_t can be modelled as a univariate ARMA process. However, if the variables are considered together, (7.26) also implies that the disturbances in the different ARMA models will be correlated and the autoregressive components will be identical. In practice though, common AR components may not be observed because of common factors in $\phi_i^*(L)$ and $\theta_i^*(L)$ in some equations.

Exercises

1. Are the following stochastic processes stationary?

(i) $y_t = \alpha \cos \lambda t + \beta \sin \lambda t$,

where $\alpha \sim N(0, \sigma^2)$, $\beta \sim N(0, \sigma^2)$ and α and β are independent;

(ii) $y_t = y_{t-1} + \epsilon_t$, $\qquad \epsilon_t \sim NID(0, \sigma^2)$;

(iii) $y_t = \epsilon_{1t} + t\epsilon_{2t}$

with

(a) $\epsilon_{it} \sim NID(\mu_i, \sigma_i^2)$, $\qquad \mu_i \neq 0, i = 1, 2$

(b) $\epsilon_{it} \sim NID(0, \sigma_i^2)$, $\qquad i = 1, 2$;

(iv) $y_t - y_{t-1} + 0.5y_{t-2} = \epsilon_t$.

2. Determine the autocorrelation function of the following stochastic process:

$$y_t = 14 + \epsilon_t + 0.4\epsilon_{t-1} - 0.2\epsilon_{t-2}.$$

3. Show that if an AR(2) process is stationary, $\rho^2(1) < [\rho(2) + 1]/2$.

4. Calculate the first five autocorrelations for an AR(2) process with

(a) $\phi_1 = 0.6, \phi_2 = -0.2$;
(b) $\phi_1 = -0.6, \phi_2 = 0.2$.

Sketch the autocorrelation functions.

5. (a) Sketch the autocorrelation function of the process

$$y_t = -0.5y_{t-1} + \epsilon_t - 0.8\epsilon_{t-1}.$$

(b) Find the coefficients, ψ_j, in the infinite MA representation.

6. Is the model

$$y_t = -0.2y_{t-1} + 0.48y_{t-2} + \epsilon_t + 0.6\epsilon_{t-1} - 0.16\epsilon_{t-2}$$

overparameterised?

7. Determine the form of the cross-correlation function between y_t and x_t in (6.6) when x_t is generated by an MA(1) process.

8. Evaluate the autocorrelation matrices at all finite lags in the vector MA(1) process

$$\begin{bmatrix} y_{1t} \\ y_{2t} \end{bmatrix} = \begin{bmatrix} \epsilon_{1t} \\ \epsilon_{2t} \end{bmatrix} + \begin{bmatrix} 0.5 & -0.3 \\ 0.7 & 0.6 \end{bmatrix} \begin{bmatrix} \epsilon_{1,t-1} \\ \epsilon_{2,t-1} \end{bmatrix}$$

Is the process invertible?

9. Consider the non-stationary processes

$$y_{1t} = y_{1,t-1} + \epsilon_{1,t} + \theta\epsilon_{1,t-1}, \qquad |\theta| < 1$$

$$y_{2t} = \phi y_{1t} + \epsilon_{2,t}, \qquad\qquad |\phi| < 1$$

Show that the vector of first differences, $(\Delta y_{1t}, \Delta y_{2t})'$, is a strictly non-invertible vector MA(1) process.

3
The Frequency Domain

1. Introduction

In the frequency domain interest is centred on the contributions made by various periodic components in the series. Such components need not be identified with regular cycles. In fact, regular cycles are unusual in economic time series, and in talking of a periodic component what is usually meant is a tendency towards cyclical movements centred around a particular frequency.

In order to gain some insight into the rather vaguely defined concept of an irregular cycle, it is first necessary to examine how regular cycles are analysed. This is done in Section 2, where it is shown how cyclical trend components may be extracted from a time series using Fourier analysis. Section 3 then establishes the link between Fourier and spectral analysis. The first step is to convert a model containing cyclical trend components to a stationary process, simply by re-interpreting the coefficients in the trend model. The second step is to permit an infinite number of cyclical components in the model. Once this is done, it is possible to give a frequency domain interpretation to the properties of the ARMA processes described in the previous chapter.

The properties of linear filters are examined in Section 5. Frequency domain analysis gives some important insights which would not be apparent in the time domain. Furthermore, the concepts introduced form the basis for cross-spectral analysis. The cross-spectrum characterises the relationship between two series in the frequency domain.

The estimation of the spectrum raises a number of issues which are not encountered in estimating the autocovariance function. These are examined in Section 6.

The Spectrum

The power spectrum of any indeterministic process of the form
(2.1.23) is defined by the continuous function

$$f(\lambda) = (2\pi)^{-1}\left[\gamma(0) + 2 \sum_{\tau=1}^{\infty} \gamma(\tau)\cos \lambda\tau\right], \qquad (1.1)$$

where λ, the frequency in radians, may take any value in the range
$[-\pi, \pi]$. However, since $f(\lambda)$ is symmetric about zero, all the
information in the power spectrum is contained in the range $[0, \pi]$.
  If y_t is white noise, $\gamma(\tau) = 0$ for $\tau \neq 0$, and so

$$f(\lambda) = \sigma^2/2\pi, \qquad (1.2)$$

where σ^2 is the variance of y_t. Thus the spectrum, which is shown in
figure 3.1, is flat. The process may be regarded as consisting of an
infinite number of cyclical components all of which have equal
weight. In fact, this effectively provides a definition of white noise in
the frequency domain.
  Looking at (1.2), it will be seen that the area under the power
spectrum over the range $[-\pi, \pi]$ is equal to the variance, σ^2. More
generally,

$$\int_{-\pi}^{\pi} f(\lambda)d\lambda = \gamma(0). \qquad (1.3)$$

This result will be demonstrated in Section 3, but for the moment it
is its interpretation which is important. Expression (1.3) shows that
the power spectrum of a linear process may be viewed as a
decomposition of the variance of the process in terms of frequency.
  The power spectrum is sometimes standardised by dividing by
$\gamma(0)$. The same effect is achieved by replacing the autocovariances in

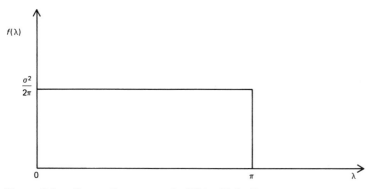

Figure 3.1   *Power Spectrum of a White Noise Process*

(1.1) by the corresponding autocorrelations. The standardised function is known as the *spectral density*, although the same term may also be used to denote the power spectrum. To add to the confusion, many authors multiply the power spectrum and spectral density by a factor of two, so that the areas under them are respectively $\gamma(0)$ and unity, when λ is defined over the range $[0, \pi]$. This usage will not be adopted here, and the discussion throughout the book will be based on the definition (1.1). The terms power spectrum and spectral density will, however, be used interchangeably, since, from the practical point of view, standardisation is not particularly important.

Defining the power spectrum over the range $[-\pi, \pi]$ when it is symmetric perhaps requires some explanation. The implicit assumption made in the discussion so far is that the series y_t is real. If y_t were complex, the power spectrum would have to be defined by the complex Fourier transform,

$$f(\lambda) = (2\pi)^{-1} \sum_{\tau=-\infty}^{\infty} \gamma(\tau) e^{-i\lambda\tau}, \qquad (1.4)$$

for λ in the range $[-\pi, \pi]$. When y_t is real, (1.4) collapses to (1.1), but this will not be true in general and $f(\lambda)$ need not be symmetric around zero. This is the reason for defining $f(\lambda)$ over the range $[-\pi, \pi]$, and although this book is restricted to real processes, it is useful to retain the more general definition for comparability with other work. Furthermore, the complex Fourier transform, (1.4), is often easier to work with when deriving theoretical results. This provides an even more important reason for its adoption.

Example 1 The MA(1) model defined in (2.1.1) has $\gamma(0) = (1 + \theta^2)\sigma^2$, $\gamma(1) = \theta\sigma^2$, and $\gamma(\tau) = 0$ for $\tau \geqslant 2$; see (2.1.14). Substituting into (1.1) gives

$$f(\lambda) = (\sigma^2/2\pi)(1 + \theta^2 + 2\theta \cos \lambda). \qquad (1.5)$$

If $\theta = 0.5$

$$f(\lambda) = \sigma^2(5 + 4 \cos \lambda)/8\pi \qquad (1.6)$$

and this is sketched in figure 3.2. Because y_t is a weighted average of current and lagged disturbance terms, the series is rather smoother than white noise. In other words, it changes 'more slowly' than white noise. In the time domain this is reflected in the positive first-order autocovariance, while in the frequency domain the same property shows up in the higher values of the power spectrum at the lower frequencies. Had the process been defined with a negative sign for θ, the spectrum would have been

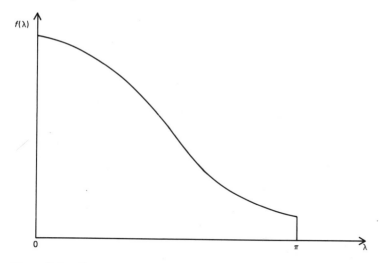

Figure 3.2 *Power Spectrum of an MA(1) Process with θ = 0.5*

greater at the higher frequencies, indicating a process more irregular than white noise.

It should be clear from the above example that the power spectrum and the autocovariance function are complementary rather than competitive. They highlight the properties of the series in different ways. The power spectrum contains no information which is not present in the autocovariance function, since it is simply a linear combination of the autocovariances.

2. Cyclical Trends

This section introduces the concept of a cyclical trend, and shows how such trends may be fitted by regression methods. A straight-forward extension of these results then leads naturally into Fourier analysis and the definition of the periodogram. In discussing this material, it is necessary to draw on certain standard results relating to trigonometric functions. These results are grouped together in Appendix B.

Cyclical Functions

The trigonometric function

$$y = \cos x \tag{2.1}$$

is defined in terms of an angle, x, which is measured in radians. Since there are 2π radians in a circle, y goes through its full complement of values as x moves from 0 to 2π. This pattern is then repeated and so for any integer, k, $\cos(x + 2k\pi) = \cos x$. The sine function exhibits a similar property, and figure 3.3 shows both $\sin x$ and $\cos x$ plotted against x. It will be observed that the cosine function is symmetric about zero. This is a reflection of the fact that it is an even function, i.e. $\cos x = \cos(-x)$. The sine function, on the other hand, is odd as $\sin x = -\sin(-x)$.

The variable y may be expressed as a cyclical function of time by defining a parameter, λ, which is measured in radians and is known as the (angular) *frequency*. The variable x is then replaced by λt. By assigning different values to λ, the function can be made to expand or contract along the horizontal axis, t. The *period* of the cycle, which is the time taken for y to go through its complete sequence of values, is equal to $2\pi/\lambda$. Thus a trigonometric function which repeated itself every five time periods would have a frequency of $2\pi/5$.

Further flexibility may be introduced into a cyclical function by multiplying the cosine, or sine, function by a parameter, ρ, known as the amplitude. Finally, there remains the problem that the position of the function is fixed with respect to the horizontal axis. This may be remedied by the introduction of an angle, θ, which is again measured in radians, and is known as the phase. Expression (2.1) therefore becomes

$$y = \rho \cos(\lambda t - \theta) = \rho \cos \lambda(t - \xi). \tag{2.2}$$

The parameter $\xi = \theta/\lambda$ gives the shift in terms of time.

Example 1 Suppose that a cosine function has a period of five,

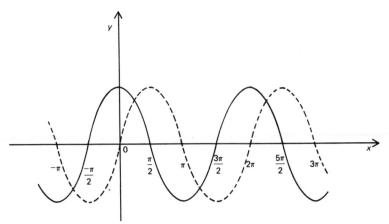

Figure 3.3 *Sine (– – –) and Cosine (——-) Functions*

but that we wish to shift it so that its peaks are at $t = 1, 7, 12, \ldots$ rather than at $t = 0, 5, 10, \ldots$. Then $\xi = 2$ and so $\theta = \lambda\xi = (2\pi/5) \cdot 2 = 4\pi/5 = 2.513$ radians. Whether this movement actually represents a forward or backward shift, however, is somewhat problematic, since the same effect could have been accomplished by setting $\xi = -3$. This would have meant $\theta = (2\pi/5)(-3) = -3.770$ radians. The ambiguity surrounding the phase has some important implications in cross-spectral analysis.

A lateral shift in a trigonometric function may be induced in a different way. Rather than introducing a phase into the sine or cosine function, it is expressed as a mixture of a sine and a cosine function. Thus (2.2) becomes

$$y = \alpha \cos \lambda t + \beta \cos \lambda t, \tag{2.3}$$

where $\alpha = \rho \cos \theta$ and $\beta = \rho \sin \theta$, and so in the example above,

$$y = \cos(\lambda t - 2.513) = -0.809 \cos \lambda t + 0.588 \sin \lambda t.$$

The transformation is often made in reverse, in which case

$$\rho^2 = \alpha^2 + \beta^2 \tag{2.4a}$$

$$\theta = \tan^{-1}(\beta/\alpha). \tag{2.4b}$$

Fourier Analysis

At the end of the previous sub-section it was observed that trigonometric terms could be imposed, one on the other, to produce a consolidated cyclical pattern. Now suppose that T points, $y_1, \ldots, y_T$, are available, and that we wish to construct a function, $y(t)$, which passes through every point. One way of achieving this objective is to let $y(t)$ be a linear function of T trigonometric terms.

If $y_1, \ldots, y_T$ have the usual interpretation as a set of time series observations, the easiest way to fit T trigonometric terms is by Fourier analysis. Let

$$n = \begin{cases} T/2 & , \quad \text{if } T \text{ is even} \\ (T-1)/2, & \quad \text{if } T \text{ is odd,} \end{cases} \tag{2.5}$$

and define the frequencies

$$\lambda_j = 2\pi j/T, \qquad j = 1, \ldots, n. \tag{2.6}$$

The first step in constructing the appropriate function, $y(t)$, is to take a pair of trigonometric terms, $\cos \lambda_j t$ and $\sin \lambda_j t$, at each of these frequencies. If T is even, there will only be one term at $j = n$ for $t = 1, \ldots, T$, since $\sin \pi t = 0$ when t is an integer. This gives exactly

$T - 1$ trigonometric terms, irrespective of whether T is odd or even. The full complement of T terms is then made up by the addition of a constant.

When T is even, the *Fourier representation* of the time series, y_t, is

$$y_t = T^{-1/2} a_0 + \sqrt{2/T} \sum_{j=1}^{n-1} (a_j \cos \lambda_j t + b_j \sin \lambda_j t)$$

$$+ T^{-1/2} a_n (-1)^t. \tag{2.7}$$

If T is odd, the last term in (2.7), which is based on $\cos \pi t$, does not appear, and the summation runs from $j = 1, \ldots, n$. Everything else remains the same. If these rules are borne in mind, the appropriate results for an odd number of observations may be obtained directly from those presented for T even. Note that in neither case is a frequency defined at less than π radians. The frequency $\lambda = \pi$ is known as the *Nyquist* frequency, and it corresponds to a period of two time units. To involve shorter periods would introduce an element of ambiguity into the proceedings, since no meaningful information on such cycles could be extracted from the observations. The introduction of the factors involving $T^{-1/2}$ is rather arbitrary, but the rationale underlying this will become clear at a later stage.

The reason for the choice of the frequencies defined by (2.6) is essentially computational. It enables the orthogonality relations in Appendix B to be exploited, thereby producing simple expressions for the Fourier coefficients, the a_j's and b_j's. Multiplying both sides of (2.7) by $\sin(2\pi i/T)$, summing from $t = 1$ to T and using (B.3b) and (B.3c) yields

$$\sum_{t=1}^{T} y_t \sin \frac{2\pi i}{T} t = (T/2)^{1/2} b_i, \qquad i = 1, \ldots, n - 1, \tag{2.8}$$

Re-arranging and setting i equal to j then gives

$$b_j = (2/T)^{1/2} \sum_{t=1}^{T} y_t \sin(2\pi j/T)t, \qquad j = 1, \ldots, n - 1. \tag{2.9a}$$

In a similar way, multiplying through by $\cos(2\pi j/T)$ produces the following expressions:

$$a_j = (2/T)^{1/2} \sum_{t=1}^{T} y_t \cos(2\pi j/T)t, \qquad j = 1, \ldots, n - 1, \tag{2.9b}$$

$$a_0 = T^{-1/2} \sum_{t=1}^{T} y_t \tag{2.9c}$$

and

$$a_n = T^{-1/2} \sum_{t=1}^{T} y_t (-1)^t. \tag{2.9d}$$

Example 2 Suppose four observations are available and that these take values $y_1 = 1$, $y_2 = 3$, $y_3 = 5$ and $y_4 = 7$. The Fourier representation is

$$y_t = T^{-1/2} a_0 + (2/T)^{1/2} a_1 \cos \frac{2\pi}{4} + (2/T)^{1/2} b_1 \sin \frac{2\pi}{4}$$

$$+ T^{-1/2} a_2 (-1)^t, \qquad t = 1, \ldots, 4. \tag{2.10}$$

Calculating the Fourier coefficients from (2.9) gives

$$y_t = 4 + 2 \cos \frac{\pi}{2} t - 2 \sin \frac{\pi}{2} t + (-1)^t$$

$$= 4 + \sqrt{8} \cos(\lambda t - \tfrac{1}{4}\pi) + (-1)^t. \tag{2.11}$$

Whether a cyclical representation is sensible in these circumstances depends on whether the observed pattern will repeat itself over subsequent sets of T observations. This implies that

$$y_{t+4k} = y_t, \qquad t = 1, 2, 3, 4, \quad k = 1, 2, \ldots$$

thereby leading to the pattern depicted in figure 3.4. Such a pattern seems rather unlikely, and faced with these data, a more reasonable course of action is simply to fit a linear trend.

Since each pair of sine and cosine terms in (2.7) can be represented

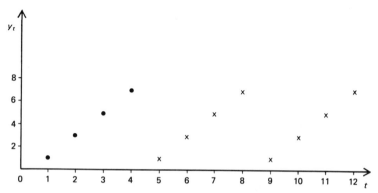

Figure 3.4   *Original Observations (●) and Pattern implied by Fourier Representation (×) in Example 2.2*

by a single translated term, the relative importance of each frequency can be measured by a single quantity. Expression (2.4a) suggests the use of

$$p_j = a_j^2 + b_j^2, \qquad j = 1, \ldots, n - 1. \tag{2.12}$$

The sum of squared deviations of the observations may be decomposed in terms of these quantities by writing

$$\sum_{t=1}^{T} (y_t - \bar{y})^2 = \sum_{j=1}^{n-1} p_j + 2p_n, \tag{2.13}$$

where $p_n = a_n^2$. This is *Parseval's theorem*. The proof, which is based on the orthogonality relations, is straightforward.

A diagram showing each p_j plotted against the corresponding period, $2\pi/\lambda_j$, is known as the *periodogram*. However, it is generally more convenient to plot p_j against λ_j. This is known as the spectrogram, although the term periodogram is also employed in this context. In fact, the most common usage of the term periodogram is for a plot of p_j against λ_j, and it is in this sense that it will be used throughout the book.

The periodogram ordinates are often defined in complex terms as

$$p_j = (2/T) \left| \sum_{t=1}^{T} y_t e^{-i\lambda_j t} \right|^2, \qquad j = 1, \ldots, n - 1. \tag{2.14}$$

However, on expanding the exponential term using the identity (A.5), imaginary quantities drop out, leaving

$$p_j = (2/T) \left[\left(\sum_{t=1}^{T} y_t \cos \lambda_j t \right)^2 + \left(\sum_{t=1}^{T} y_t \sin \lambda_j t \right)^2 \right]. \tag{2.15}$$

This expression follows immediately from the formulae in (2.9).

Equations (2.14) and (2.15) may be written with y_t replaced by $(y_t - \bar{y})$. An alternative form for (2.14) is therefore

$$p_j = \frac{2}{T} \left| \sum_{t=1}^{T} (y_t - \bar{y}) e^{-i\lambda_j t} \right|^2, \qquad j = 1, \ldots, n - 1. \tag{2.16}$$

This can be verified directly by noting that

$$\sum_{t=1}^{T} (y_t - \bar{y}) e^{-i\lambda_j t} = \sum_{t=1}^{T} y_t e^{-i\lambda_j t} - \bar{y} \sum_{t=1}^{T} e^{-i\lambda_j t},$$

and observing that the second term is zero in view of the identity in (B.5). For $j = 0$, the use of formula (2.16) leads to $p_0 = 0$. However, this makes sense, since otherwise p_0 is a reflection of the mean and this is of little interest in the context of studying the cyclical movements in a series.

Cyclical Trend Models

The value of the Fourier representation, (2.7), can only be assessed by relating it to some underlying model. Consider the cyclical trend model,

$$y_t = T^{-1/2}\alpha_0 + \sqrt{2/T} \sum_{j=1}^{n-1} (\alpha_j \cos \lambda_j t + \beta_j \sin \lambda_j t)$$

$$+ T^{-1/2}\alpha_n(-1)^t + \epsilon_t, \qquad t = 1, \ldots, T, \tag{2.17}$$

where $\epsilon_t \sim NID(0, \sigma^2)$. Within this framework, the Fourier coefficients, (2.9), are seen to be the OLS estimators of the parameters in (2.17). By defining the $T \times 1$ vector, $\gamma = (\alpha_0, \alpha_1, \beta_1, \alpha_2, \ldots, \alpha_{T/2})'$, (2.17) may be written in matrix notation as

$$y = Z\gamma + \epsilon, \tag{2.18}$$

where Z is the $T \times T$ *Fourier matrix*, shown overleaf in (2.19). It follows from (B.3) that Z is an orthogonal matrix, and so the OLS estimator of γ is given by

$$c = (Z'Z)^{-1}Z'y = Z'y. \tag{2.20}$$

This is exactly the result in (2.9). Because of the orthogonality of the regressors, the individual OLS estimators are independent of each other. Explanatory variables may therefore be removed from (2.17) without affecting the estimates of the parameters which remain.

The distributional properties of the Fourier coefficients may be derived immediately from standard least squares theory. Thus

$$\text{Var}(c) \sim \sigma^2(Z'Z)^{-1} = \sigma^2 I, \tag{2.21}$$

and so the elements of $c - \gamma$ are normally and independently distributed with zero mean and variance σ^2. It follows that when $\alpha_j = \beta_j = 0$,

$$\sigma^{-2}p_j \sim \chi_2^2, \qquad j = 1, \ldots, n-1. \tag{2.22}$$

Unless the model is fitted with the full complement of T regressors, an estimate of σ^2 can be computed from the residuals. The usual formula for s^2 is applicable, and so the statistic $s^{-2}p_j/2$ may be tested against an F-distribution. The interpretation of the null hypothesis is that there is no component of frequency λ_j in the model.

$$
Z =
\begin{bmatrix}
\dfrac{1}{\sqrt{T}} & \sqrt{\dfrac{2}{T}}\cdot\cos\dfrac{2\pi}{T} & \sqrt{\dfrac{2}{T}}\cdot\sin\dfrac{2\pi}{T} & \sqrt{\dfrac{2}{T}}\cdot\cos\dfrac{4\pi}{T} & \cdots & -\sqrt{\dfrac{1}{T}} \\[2ex]
\dfrac{1}{\sqrt{T}} & \sqrt{\dfrac{2}{T}}\cdot\cos\dfrac{2\pi}{T}2 & \sqrt{\dfrac{2}{T}}\cdot\sin\dfrac{2\pi}{T}2 & \sqrt{\dfrac{2}{T}}\cdot\cos\dfrac{4\pi}{T}2 & \cdots & \sqrt{\dfrac{1}{T}} \\[2ex]
\cdots & \cdots & \cdots & \cdots & \cdots & \cdots \\[2ex]
\dfrac{1}{\sqrt{T}} & \sqrt{\dfrac{2}{T}}\cdot\cos\dfrac{2\pi}{T}(T-1) & \sqrt{\dfrac{2}{T}}\cdot\sin\dfrac{2\pi}{T}(T-1) & \sqrt{\dfrac{2}{T}}\cdot\cos\dfrac{4\pi}{T}(T-1) & \cdots & -\sqrt{\dfrac{1}{T}} \\[2ex]
\dfrac{1}{\sqrt{T}} & \sqrt{\dfrac{2}{T}} & 0 & \sqrt{\dfrac{2}{T}} & \cdots & \sqrt{\dfrac{1}{T}}
\end{bmatrix}
\tag{2.19}
$$

Example 3   Anderson (1971, pp. 106–7) uses a cyclical trend plus error model to capture the seasonal pattern in a three year time series of monthly observations. The *fundamental* seasonal frequency for monthly data is $\lambda = 2\pi/12$, corresponding to a period of 12 months. The remaining frequencies needed to pick up the full seasonal pattern are at $\lambda_j = 2\pi j/12$, $j = 2, \ldots, 6$. These are known as the *harmonic* frequencies. The (partial) periodogram is shown in figure 3.5. The two lowest frequency components, corresponding to periods of 12 and 6 months, account for most of the seasonal pattern. The remaining components are relatively unimportant and are not, in fact, statistically significant at the 5% level.

A given seasonal pattern may also be represented by a constant term and eleven 'dummy' variables. However, as the above example illustrates, a seasonal pattern can often be represented very closely by only a limited number of trigonometric terms. Omitting some of the dummy variables, on the other hand, is likely to be a good deal less satisfactory.

In the discussion so far it has been assumed that all the frequencies to be included in a cyclical trend model are contained within the set defined by (2.6). There is no reason why this should necessarily be the case. In general, a cyclical trend plus error model may be written

$$y_t = \alpha_0 + \sum_j (\alpha_j \cos \lambda_j t + \beta_j \sin \lambda_j t) + \epsilon_t, \tag{2.23}$$

where $0 < \lambda_j \leqslant \pi$ for all j. The parameters in the model may still be estimated by OLS, but unless the frequencies are all contained within the set defined by (2.6), the regressors will no longer be orthogonal.

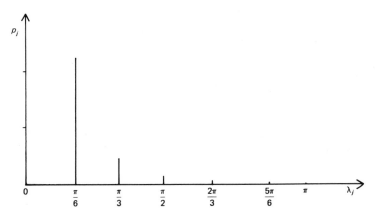

Figure 3.5   *Periodogram Ordinates for Seasonal Frequencies in Example 2.3*

3. Spectral Representation of a Stochastic Process

Consider the model

$$y_t = \alpha \cos \lambda t + \beta \sin \lambda t, \qquad t = 1, \ldots, T, \tag{3.1}$$

where α and β are fixed parameters and λ takes a particular value between 0 and π. This can be regarded as one of the deterministic components in a cyclical trend model of the form (2.17). It is clearly not stationary, as $E(y_t)$ is a function of t, but this state of affairs can be remedied by replacing α and β by two *random variables*, u and v, which are uncorrelated and have mean zero and variance σ^2. It then follows that

$$E(y_t) = E(u)\cos \lambda t + E(v)\sin \lambda t = 0, \tag{3.2a}$$

$$\text{Var}(y_t) = E[(u \cos \lambda t + v \sin \lambda t)^2]$$
$$= \sigma^2(\cos^2 \lambda t + \sin^2 \lambda t) = \sigma^2, \tag{3.2b}$$

and

$$\gamma(\tau) = E(y_t y_{t-\tau}) = \sigma^2[\cos \lambda t \cos \lambda(t - \tau) + \sin \lambda t \sin \lambda(t - \tau)]$$
$$= \sigma^2 \cos \lambda\tau, \qquad \tau = 1, 2, 3, \ldots \tag{3.2c}$$

Although u and v are random variables, they are *fixed* in any particular realisation. Thus, although it is stationary, the model is still deterministic; only two observations are necessary to determine u and v exactly, and once this has been done the remaining points in the series can be forecast with zero mean square error. In practice, therefore, the only difference between the non-stationary model (3.1) and the corresponding stationary model is in the interpretation of the parameters.

The stationary model can be extended to include terms at more than one frequency. If J frequencies, $\lambda_1, \ldots, \lambda_J$, are included then

$$y_t = \sum_{j=1}^{J} (u_j \cos \lambda_j t + v_j \sin \lambda_j t). \tag{3.3}$$

If

$$E(u_j) = E(v_j) = 0, \qquad j = 1, \ldots, J, \tag{3.4a}$$

$$\text{Var}(u_j) = \text{Var}(v_j) = \sigma_j^2, \qquad j = 1, \ldots, J, \tag{3.4b}$$

$$E(u_i u_j) = E(v_i v_j) = 0, \qquad i \neq j \tag{3.4c}$$

and

$$E(u_i v_j) = 0, \qquad \text{for all } i, j, \tag{3.4d}$$

the mean of the series is zero while

$$\gamma(0) = \sum_{j=1}^{J} \sigma_j^2 \tag{3.5}$$

and

$$\gamma(\tau) = \sum_{j=1}^{J} \sigma_j^2 \cos \lambda\tau. \tag{3.6}$$

Expression (3.5) shows that the variance of the process can be regarded as the sum of the variances of the u_j's and v_j's. If the set of frequencies in (3.3) corresponded to the set of frequencies defined by (2.6), this result would be similar to (2.13), but with the important difference that in (2.13) it is the variance *in the sample* which is being decomposed into estimates of fixed quantities, namely the amplitudes of the various trigonometric components.

Now consider the white noise process, defined by (2.1.12). Since this is completely random, all of its frequency components are of equal importance, but this immediately raises the question of what is meant by 'all' in this context. It makes no sense to consider a particular set of frequencies and say that they all make an equal contribution to the variance of the process, since there are an infinite number of frequencies in the range 0 to π. In order to extend equation (3.3) to cover this type of situation, it is necessary to allow an infinite number of trigonometric terms to be included in the range 0 to π. This is achieved by letting $J \to \infty$ and replacing the summation sign by an integral. The u_j's and v_j's are then replaced by *continuous* functions of λ denoted by $u(\lambda)$ and $v(\lambda)$ and the resulting *spectral representation* is given by

$$y_t = \int_0^\pi u(\lambda)\cos \lambda t \, d\lambda + \int_0^\pi v(\lambda)\sin \lambda t \, d\lambda. \tag{3.7}$$

This is often referred to as *Cramér's representation*.

The functions $u(\lambda)$ and $v(\lambda)$ are stochastic, with properties somewhat analogous to those of the u_j's and v_j's in the discrete representation, (3.3). Consider any small interval, $d\lambda$, and let

$$E[u(\lambda)d\lambda] = E[v(\lambda)d\lambda] = 0 \tag{3.8a}$$

for all such intervals in the range $[0, \pi]$. If this assumption holds, applying the expectation operator to (3.7) yields $E(y_t) = 0$. In a similar way let

$$\text{Var}[u(\lambda)d\lambda] = \text{Var}[v(\lambda)d\lambda] = f(\lambda)d\lambda \tag{3.8b}$$

where $f(\lambda)$ is a continuous function, and suppose that

$$E[u(\lambda_1)d\lambda_1 u(\lambda_2)d\lambda_2] = E[v(\lambda_1)d\lambda_1 v(\lambda_2)d\lambda_2] = 0, \qquad (3.8c)$$

where $d\lambda_1$ and $d\lambda_2$ are non-overlapping intervals, centred on λ_1 and λ_2 respectively, and (3.8c) is true for all possible non-overlapping intervals in the range $[0, \pi]$. Finally let

$$E[u(\lambda)d\lambda_1 v(\lambda)d\lambda_2] = 0 \qquad (3.8d)$$

for all intervals, including cases where $d\lambda_1 = d\lambda_2$. From the spectral representation, (3.7),

$$\gamma(\tau) = E(y_t y_{t-\tau}) = E\left\{ \left[\int_0^\pi u(\lambda)\cos \lambda t \, d\lambda + v(\lambda)\sin \lambda t \, d\lambda \right] \right.$$
$$\left. \times \left[\int_0^\pi u(\lambda)\cos \lambda(t-\tau)d\lambda + v(\lambda)\sin \lambda(t-\tau)d\lambda \right] \right\} \quad (3.9)$$

but in view of assumption (3.8d) all cross product terms involving sines and cosines disappear on expanding this expression out. Taking expectations of the remaining terms and using the trigonometric addition formulae (A.1) and (A.2) yields

$$\gamma(\tau) = 2\int_0^\pi f(\lambda)\cos \lambda\tau \, d\lambda, \qquad \tau = 0, \pm 1, \pm 2, \ldots \qquad (3.10)$$

All indeterministic processes have a spectral representation. Setting $\tau = 0$ in (3.10) gives

$$\gamma(0) = 2\int_0^\pi f(\lambda)d\lambda, \qquad (3.11)$$

showing that the power spectrum may be interpreted as a decomposition of the variance into cyclical components over the continuum $[0, \pi]$. In order to determine $f(\lambda)$ from the auto-covariances, a Fourier transform of (3.10) is taken; see Appendix C. This gives the formula stated in (1.1).

Complex Spectral Representation*

The complex spectral representation is more compact than (3.7) and easier to handle mathematically. Let i be defined such that $i^2 = -1$. For a purely indeterministic process

$$y_t = \int_{-\pi}^\pi e^{i\lambda t} z(\lambda)d\lambda, \qquad (3.12)$$

where for any small interval, $d\lambda$, in the range $-\pi \leqslant \lambda \leqslant \pi$

$$E[z(\lambda)d\lambda] = 0 \tag{3.13a}$$

and

$$E[z(\lambda)d\lambda \cdot \overline{z(\lambda)d\lambda}] = f(\lambda)d\lambda, \tag{3.13b}$$

while for any two non-overlapping intervals, $d\lambda_1$ and $d\lambda_2$, centred on λ_1 and λ_2 respectively,

$$E[z(\lambda_1)d\lambda_1 \cdot \overline{z(\lambda_2)d\lambda_2}] = 0. \tag{3.13c}$$

These properties are analogous to those set out in (3.8). As before, it follows almost immediately that $E(y_t) = 0$ for all t, while

$$\gamma(\tau) = E[y_t\overline{y_{t-\tau}}] = E\left[\int_{-\pi}^{\pi} e^{i\lambda t} z(\lambda)d\lambda \cdot \int_{-\pi}^{\pi} e^{-i\lambda(t-\tau)}\overline{z(\lambda)d\lambda} \right].$$

In view of (3.13b) and (3.13c), this becomes

$$\gamma(\tau) = \int_{-\pi}^{\pi} e^{i\lambda\tau} f(\lambda)d\lambda. \tag{3.14}$$

The inverse Fourier transform of (3.14) is (1.4).

Wold Decomposition Theorem*

In moving from (3.3) to (3.7), a strong distinction was drawn between deterministic and indeterministic processes. However, it is possible to conceive of a process containing both deterministic and indeterministic components. In fact, a general result known as the *Wold decomposition theorem* states that any covariance stationary process can be represented as

$$y_t = y_t^* + y_t^\dagger,$$

where y_t^* is a deterministic component, as in the right-hand side of (3.3), and $y_t^\dagger$ is a linear process as defined by (2.1.23). Since the spectrum of y_t^* is discrete, the frequency domain properties of the process as a whole must be described in terms of the *spectral distribution function*; this shows a jump at each of the frequencies, λ_j, appearing in (3.3).

Perfectly regular periodic components are unusual in economic time series. To the extent that such a component were present in a series, it would tend to show up as a sharp spike in the estimated spectrum. Irregular cycles emerge as broad peaks. The spectrum for a series with an approximate five year cycle, shown later in figure 3.8(b) on page 79, illustrates this point.

4. Properties of Autoregressive-Moving Average Processes in the Frequency Domain

In the time domain, the properties of ARMA processes are characterised by their autocovariance, or equivalently autocorrelation, functions. However, while the autocovariance function may highlight certain features of the process, there may be other aspects which emerge less clearly. In particular, the cyclical properties of the series may not be well characterised by the autocovariances, especially if this behaviour is at all complex. For this reason, it is often desirable to examine the power spectrum of the process, as well as the autocovariance function.

Consider a stationary series, x_t, generated by an indeterministic stochastic process. Suppose that a new series, y_t, is constructed as a weighted average of $x_{t+s}, \ldots, x_t, \ldots, x_{t-r}$, where the weights $w_{t+s}, \ldots, w_t, \ldots, w_{t-r}$ are real and satisfy the condition $\Sigma w_j^2 < \infty$. This may be written as

$$y_t = \sum_{j=-r}^{s} w_j x_{t-j}. \tag{4.1}$$

The operation in (4.1) is termed a *linear time invariant* filter. Time invariance in this context simply means that the weights are independent of t. When the weights sum to one, (4.1) is sometimes referred to as a 'moving average', but this term should obviously be used with some care in order to avoid confusion with a moving average *process*.

The relevance of (4.1) is that it serves to introduce a result which forms the basis for obtaining the power spectrum of any ARMA process. Furthermore, in the next section, the same result is used to study the effect of various filtering operations in the frequency domain. This enables some insight to be gained into certain operations which are typically carried out on time series, for example to seasonally adjust or smooth them.

Let $f_x(\lambda)$ denote the spectrum of x_t and define

$$W(\lambda) = \sum_{j=-r}^{s} w_j e^{-i\lambda j}. \tag{4.2}$$

The spectrum of y_t is then given by

$$f_y(\lambda) = |W(\lambda)|^2 f_x(\lambda), \tag{4.3}$$

the factor $|W(\lambda)|^2$ being known as the *transfer function*.

The reasoning behind this result is set out in the next section, but before proceeding further, its importance will be illustrated with respect to the MA(1) process, (2.1.1). In terms of (4.1), $w_0 = 1$ and $w_1 = \theta$, all other weights being zero, and so

$$W(\lambda) = 1 + \theta e^{-i\lambda}. \tag{4.4}$$

Therefore

$$
\begin{aligned}
|W(\lambda)|^2 &= W(\lambda)\overline{W(\lambda)} \\
&= (1 + \theta e^{-i\lambda})(1 + \theta e^{i\lambda}) \\
&= 1 + \theta^2 + \theta(e^{-i\lambda} + e^{i\lambda}) \\
&= 1 + \theta^2 + 2\theta \cos \lambda.
\end{aligned}
$$

Since x_t is white noise in this case, $f_x(\lambda) = \sigma^2/2\pi$. The spectrum of y_t is therefore

$$f_y(\lambda) = (\sigma^2/2\pi)(1 + \theta^2 + 2\theta \cos \lambda). \tag{4.5}$$

This is identical to formula (1.5), which was obtained by applying a Fourier transform to the autocovariances. The importance of (4.3) is that it makes it unnecessary to derive the autocovariance function of y.

*Derivation of the Transfer Function**

The spectral representation of x_t in complex terms is

$$x_t = \int_{-\pi}^{\pi} e^{i\lambda t} z_x(\lambda) d\lambda, \tag{4.6}$$

where

$$E[z_x(\lambda_1)d\lambda_1 \cdot \overline{z_x(\lambda_2)d\lambda_2}] = \begin{cases} f_x(\lambda)d\lambda, & \text{for } \lambda_1 = \lambda_2 \\ 0, & \text{for } \lambda_1 \neq \lambda_2. \end{cases}$$

Substituting the spectral representation of x_t in (4.1) gives

$$
\begin{aligned}
y_t &= \sum_{-r}^{s} w_j \int_{-\pi}^{\pi} e^{i(t-j)\lambda} z_x(\lambda) d\lambda \\
&= \int_{-\pi}^{\pi} e^{i\lambda t} (\Sigma w_j e^{-i\lambda j}) z_x(\lambda) d\lambda \\
&= \int_{-\pi}^{\pi} e^{i\lambda t} W(\lambda) z_x(\lambda) d\lambda. \tag{4.7}
\end{aligned}
$$

Expression (4.7) is the spectral representation of y_t, with

$z_y(\lambda) = W(\lambda)z_x(\lambda)$. Thus, for a given value of λ, $z_x(\lambda)$ is multiplied by $W(\lambda)$, the *frequency response function*. By definition,

$$f_y(\lambda)d\lambda = E[z_y(\lambda)d\lambda \cdot \overline{z_y(\lambda)d\lambda}]$$

and so (4.3) follows immediately, given the definition of $f_x(\lambda)$.

Autoregressive Processes

A stationary AR(1) process may be expressed as an infinite moving average as in (2.2.5). Therefore

$$W(\lambda) = \sum_{j=0}^{\infty} \phi^j e^{-i\lambda j}$$

$$= \sum_{j=0}^{\infty} (\phi e^{-i\lambda})^j$$

$$= 1/(1 - \phi e^{-i\lambda}),$$

the summation being possible since $e^{-i\lambda} \leqslant 1$ for all λ. Hence the spectrum of y_t is

$$f_y(\lambda) = \left(\frac{\sigma^2}{2\pi}\right)\left(\frac{1}{|1 - \phi e^{-i\lambda}|^2}\right) = \left(\frac{\sigma^2}{2\pi}\right)\left(\frac{1}{1 + \phi^2 - 2\phi \cos \lambda}\right). \quad (4.8)$$

An alternative approach, based on the knowledge that y_t does indeed have a continuous spectrum, is to write

$$\epsilon_t = y_t - \phi y_{t-1},$$

and proceed much as in the MA(1) case to derive

$$f_\epsilon(\lambda) = |1 - \phi e^{-i\lambda}|^2 \cdot f_y(\lambda).$$

Expression (4.8) then follows immediately on setting $f_\epsilon(\lambda) = \sigma^2/2\pi$ and re-arranging.

The above method of derivation is easy to apply to higher order processes. For an AR(p) model,

$$W(\lambda) = 1\bigg/\left(1 - \sum_{j=1}^{p} \phi_j e^{-i\lambda j}\right). \quad (4.9)$$

The spectrum of an AR(1) process with $\phi > 0$ is similar to the MA(1) spectrum shown in figure 3.2. The positive serial correlation is reflected in the high values of $f(\lambda)$ at the low frequencies. As ϕ tends towards one, the contribution of these lower frequencies to the variance of the process increases as the series becomes more 'slowly changing'.

For the AR(2) process, the spectrum is given by

$$f_y(\lambda) = \left(\frac{\sigma^2}{2\pi}\right)\left(\frac{1}{1 + \phi_1^2 + \phi_2^2 - 2\phi_1(1 - \phi_2)\cos\lambda - 2\phi_2\cos 2\lambda}\right).$$
(4.10)

The shape of the spectrum depends crucially on the values taken by the parameters. A spectrum in which $f(\lambda)$ monotonically decreases or increases could obviously arise as the AR(1) model is a special case of AR(2). However, in Section 2.2, attention was drawn to AR(2) processes in which the roots of the characteristic equation were complex. This led to a damped cyclical pattern in the autocorrelation function and it was pointed out that this was an indication of some kind of cyclical behaviour in the series. The actual nature of these movements emerges much more clearly in the frequency domain. Figure 3.6 shows the spectrum for an AR(2) process in which $\phi_1 = 0.7$ and $\phi_2 = -0.5$. The peak indicates a tendency towards a cycle at a frequency around λ^*. This may be termed *pseudo-cyclical* behaviour, since the movements are not regular. A deterministic cycle, say of the form (3.1), would show up as a sharp spike.

When the spectrum contains a peak within the range $0 < \lambda < \pi$, its exact position may be determined by setting the derivative of (4.10) equal to zero. Solving the resulting equation yields

$$\lambda_0^* = \cos^{-1}[-\phi_1(1 - \phi_2)/4\phi_2].$$
(4.11)

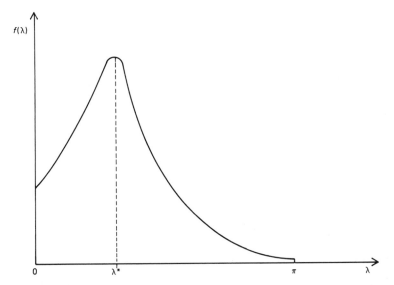

Figure 3.6 *Power Spectrum of an AR(2) Process with $\phi_1 = 0.7$ and $\phi_2 = -0.5$*

For the process in figure 3.6, $\lambda_0^* = \cos^{-1}(0.5250) = 1.018$, which coincides with a period of 6.172. This is not the same as the period of the autocorrelation function, which is determined from (2.2.15), although the two figures are reasonably close. On a related point, it is not true to say that complex roots necessarily imply a peak in the spectrum in the range $0 < \lambda < \pi$, although this will generally be the case.

The type of pseudo-cyclical behaviour exhibited by the AR(2) process is attractive from the point of view of economic modelling. Seasonal movements aside, regular cyclical patterns are unlikely to occur in economics, yet series generated by AR(2) processes can easily exhibit the kind of fluctuations often observed in practice.

Mixed Processes and the Associated Polynomials

Using the techniques already developed, it is not difficult to show that the spectrum of an ARMA(p, q) process is given by

$$f_y(\lambda) = \frac{\sigma^2}{2\pi} \cdot \frac{\left| 1 + \sum_{j=1}^{q} \theta_j e^{-i\lambda j} \right|^2}{\left| 1 - \sum_{j=1}^{p} \phi_j e^{-i\lambda j} \right|^2}. \tag{4.12}$$

This is known as a *rational spectrum*.

It is worth noting that there is a very convenient link between the frequency response function and the associated polynomials defined in (2.1.26). In the MA(1) process, for example, the associated polynomial is

$$\theta(L) = 1 + \theta L,$$

but replacing the lag operator by $e^{-i\lambda}$ gives the frequency response function, (4.4). More generally, if an ARMA(p, q) process is written in terms of associated polynomials, the spectrum is given immediately by

$$f_y(\lambda) = \frac{\sigma^2}{2\pi} \cdot \frac{|\theta(e^{-i\lambda})|^2}{|\phi(e^{-i\lambda})|^2}. \tag{4.13}$$

5. Properties of Linear Filters

The concept of a linear filter was introduced in the previous section and defined in equation (4.1). Adjusting data by applying linear filters or 'moving averages' is very common, particularly with economic time series. Such filters may be applied for a number of reasons. At the simplest level is the idea that a time series consists of a trend, upon which is superimposed a random disturbance term. The trend is regarded as a smooth function of time, but in view of the disturbance term the graph of observed points is rather irregular. This leads to the idea of *smoothing*, a technique by which the trend at a given point in time is represented by a weighted average of the observations around that point.

As an example of smoothing, consider a five period moving average

$$y_t = \frac{1}{5} \sum_{j=-2}^{2} x_{t-j}, \qquad t = 3, \ldots, T - 2. \tag{5.1}$$

The weights in (5.1) are equal and sum to unity. The irregularity of the disturbance term means that there will be some cancelling out over the five periods. The effect of the disturbances will therefore be muted, although obviously not entirely eliminated.

The use of (5.1) to represent the trend raises questions concerning the distortions which might arise from using observations at times other than t in the filter. Commonsense would suggest that more weight be given to x_t, and the values closest to it. Indeed, some rationalisation can be given for this by assuming that the trend can be represented by a low order polynomial in the locality of t. For a five period filter, this yields the formula

$$y_t = (-3x_{t-2} + 12x_{t-1} + 17x_t + 12x_{t+1} - 3x_{t+2})/35. \tag{5.2}$$

Further details will be found in Anderson (1971, Section 3.3).

One motive for applying a linear filter to a time series is to mitigate the effect of randomness in the data. Another is to remove a cyclical component. The effect of (4.1) with annual data, for example, would be to smooth out a five year cycle from the data. Exactly what is meant by saying that a cyclical component is 'removed' or 'smoothed out', however, only becomes clear when the filter itself is examined in the frequency domain. As well as giving more precise meaning to these concepts, such analysis also enables some comparison to be made between different procedures. A five year cycle could also be treated by a differencing operation of the form

$$y_t = x_t - x_{t-5}, \qquad t = 6, \ldots, T, \tag{5.3}$$

and it is interesting to contrast the effect this has on the series with that produced by (5.1).

The treatment of seasonality has an important role, particularly in economic time series. Examining the effect of linear filters can give a good deal of insight into some of the actual and potential effects of seasonal adjustment. Unfortunately, the methods adopted by government statistical offices with regard to seasonal adjustment frequently imply some kind of non-linear filtering operation. The methods discussed below cannot therefore be applied directly, although the concepts introduced can be used to compare the adjusted and unadjusted series by cross-spectral analysis.

Gain and Phase

The effect of a linear filter on a series is twofold. The first effect is to change the relative importance of the various cyclical components. This becomes clear in an examination of the spectrum of the process before and after the filter is applied. In the previous section it was shown how this effect is captured in the transfer function, the form of which is easily derived once the filter is given. The study of the transfer function also plays an important role in this section, although it will often be more convenient to work with its square root, $|W(\lambda)|$. This quantity, which will be denoted by $G(\lambda)$, is termed the *gain*.

The other property of a filter concerns the extent to which it induces a shift in the series with regard to its position in time. Suppose that

$$y_t = x_{t-3}. \tag{5.4}$$

The effect of this rather trivial filtering operation is to shift the series back by three time periods. If x_t were generated by a cyclical process, say

$$x_t = \cos \lambda t, \tag{5.5}$$

the filtered series would be

$$y_t = \cos \lambda(t - 3) = \cos(\lambda t - 3\lambda). \tag{5.6}$$

The *phase* shift, in radians, is given by $Ph(\lambda) = 3\lambda$. Thus a lateral shift of three periods in the time domain becomes a phase change of 3λ in the frequency domain.

A graph of $Ph(\lambda)$ against λ is termed the *phase diagram*. For the pure delay given in (5.4), $Ph(\lambda)$ is a straight line through the origin with a slope equal to the time lag; see figure 3.7.

The phase may be obtained from the frequency response function. If $W(\lambda) = W^*(\lambda) + iW^\dagger(\lambda)$, where $W^*(\lambda)$ are both real, then

$$Ph(\lambda) = \tan^{-1}[-W^\dagger(\lambda)/W^*(\lambda)], \qquad 0 \leqslant \lambda \leqslant \pi. \tag{5.7}$$

The rationale behind this formula is best explained by reference to

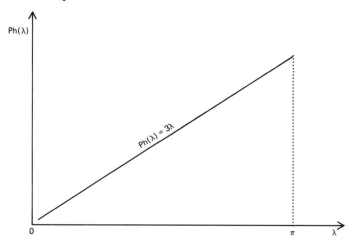

Figure 3.7 *Phase Diagram for $y_t = x_{t-3}$*

the Cramér representation of a stochastic process. This is discussed in the next (starred) sub-section. For a given Ph(λ), the shift in time units is Ph(λ)/λ. Except when Ph(λ) is a linear function of λ, as in the example above, the time shift will be different for different values of λ.

The definition of the phase is subject to a fundamental ambiguity which can sometimes give rise to problems. This is because adding or subtracting a complete cycle from an angle will not change its tangent. Thus if a particular angle, denoted, say, by Ph*(λ), emerges as a solution to (5.7), any angle satisfying the relationship Ph(λ) = Ph*(λ) $\pm$ $2h\pi$, where h is an integer, is also a valid solution of (5.7). In the case of the simple filter, (5.4), the general solution is Ph(λ) = 3λ $\pm$ $2h\pi$. Therefore, at a frequency of $\lambda = \pi/2$, for example, the phase could be interpreted not as a lag of 3, but as a lead of $-3 + 2\pi/(\pi/2) = 1$. However, taking a solution other than Ph(λ) = 3λ in this case would result in a non-linear phase function which would not have the interpretation of a constant lag in the time domain.

The Frequency Response Function*

The frequency response function will, in general, be complex. It is convenient to write it in polar form,

$$W(\lambda) = G(\lambda)e^{-i\text{Ph}(\lambda)}, \tag{5.8}$$

where $G(\lambda) = |W(\lambda)|$ and Ph(λ) is as defined in (5.7). The spectral representation of a process obtained by linear filtering was derived in the previous section, and is given by (4.7). Substituting for $W(\lambda)$

from (5.8) yields

$$y_t = \int_{-\pi}^{\pi} e^{i\lambda t} G(\lambda) e^{-i\mathrm{Ph}(\lambda)} z_x(\lambda) d\lambda$$

$$= \int_{-\pi}^{\pi} e^{i\lambda(t - \mathrm{Ph}(\lambda)/\lambda)} G(\lambda) z_x(\lambda) d\lambda. \tag{5.9}$$

The interpretation of the gain and phase should now be clear. At each frequency the amplitude component, $z_x(\lambda)$, is multiplied by $G(\lambda)$, while the nature of the phase shift emerges in the term $(t - \mathrm{Ph}(\lambda)/\lambda)$, i.e. for periodic components of frequency λ, y lags x by $\mathrm{Ph}(\lambda)/\lambda$ periods.

Smoothing and the Removal of Cycles

The frequency response function of the linear filter in (5.1) is given by

$$W(\lambda) = \frac{1}{5} \sum_{j=-2}^{2} e^{-i\lambda j}. \tag{5.10}$$

Using (A.9) and (A.10)

$$W(\lambda) = \frac{\sin(5\lambda/2)}{5 \sin(\lambda/2)} \tag{5.11}$$

and so $W(\lambda)$ is real in this case. This reflects a more general result, namely that a filter which is symmetric around t does not exhibit a phase shift. The result follows because the sine function is odd and so

$$\sum_{j=-s}^{s} w_j e^{-i\lambda j} = \sum_{j=-s}^{s} w_j \cos \lambda j - i \sum_{j=-s}^{s} w_j \sin \lambda j$$

$$= w_0 + 2 \sum_{j=1}^{s} w_j \cos \lambda j. \tag{5.12}$$

Because (5.10) is real, the gain is simply its absolute value. Both the smoothing of the series and the effect on the cycle can be seen in figure 3.8(a). The irregularity in the series is dampened down because the gain at high frequencies is very small. Thus the spectrum of the filtered series will have relatively little power for frequencies close to π, reflecting the fact that it is now more slowly changing than before.

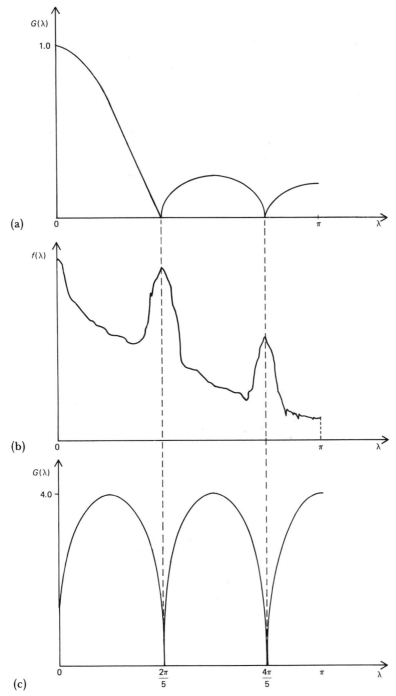

Figure 3.8 *Graphs showing: (a) the Gain for a Five Year Moving Average with Annual Data; (b) the Spectrum of a Series with an Irregular Five Year Cycle; and (c) the Gain for a Five Year Differencing Filter*

If a five year cycle were present in the data, the spectrum of the original series might be something like that shown in figure 3.8(b). The peak at $\lambda = 2\pi/5$ corresponds to a period of five years. This is the *fundamental frequency*. The second peak, at $\lambda = 4\pi/5$, is the first *harmonic*. No further harmonics are shown in the diagram, since the second harmonic is at $6\pi/5$ and this is below the Nyquist frequency, π. The effect of the filter is to eliminate completely the cyclical components at $\lambda = 2\pi/5$ and $\lambda = 4\pi/5$, since at these points the gain is zero. However, because the cycle is not regular, some of the power around the fundamental and harmonic frequencies will remain.

The differencing filter, (5.3), provides an alternative way of handling the five year cycle. The frequency response function in this case is

$$W(\lambda) = 1 - e^{-i5\lambda} \tag{5.13}$$

and so

$$G(\lambda) = (1 - e^{-i5\lambda})(1 - e^{i5\lambda})$$

$$= 1 - e^{-i5\lambda} - e^{i5\lambda} + 1$$

$$= 2(1 - \cos 5\lambda). \tag{5.14}$$

The graph of the gain, shown in figure 3.8(c), provides an interesting contrast to the corresponding graph for the moving average filter. In both cases the gain is zero at the fundamental frequency and its harmonic, but while the moving average filter removes a good deal of power from the high frequencies, the differencing filter has the opposite effect. In particular, the gain at $\lambda = 0$ is zero. The moving average filter therefore smooths out the irregular high frequency components, leaving what may be identified as the trend, while the differencing operation essentially removes the trend.

Unlike (5.1), the differencing filter also has a phase effect. This is given by

$$Ph(\lambda) = \tan^{-1}[(-\sin 5\lambda)/(1 - \cos 5\lambda)]$$

and the shift induced by the filter at different frequencies may be examined by constructing the corresponding phase diagram.

Seasonal Adjustment

Suppose a series consists of monthly data. This may be seasonally adjusted by applying a linear filter with weights

$$w_j = \begin{cases} 1/12, & j = 0, \pm 1, \pm 2, \ldots, \pm 5, \\ 1/24, & j = \pm 6. \end{cases} \tag{5.15}$$

The frequency response function of the seasonal adjustment filter is

$$W(\lambda) = \frac{\sin 6\lambda \cos(\lambda/2)}{12 \sin(\lambda/2)}. \tag{5.16}$$

This expression may be derived in much the same way that (5.11) was obtained from (5.10), although because of the unequal weights the algebra is a little more complicated. As might be expected from the discussion in the preceding section, the seasonal adjustment filter has a gain of zero at the fundamental seasonal frequency, $\pi/6$, and at the harmonics $\lambda_j = 2\pi j/12, j = 2, \ldots, 6$. Thus the power at these frequencies is eliminated. The other effect of the filter is to smooth the series by attenuating the higher frequencies; see Fishman (1969, p. 46).

Applying the filter defined by (5.15) is a relatively simple way of carrying out seasonal adjustment. Nevertheless, an analysis of its properties raises some fundamental issues regarding the nature of seasonal adjustment procedures in general, and what they are designed to achieve. The spectrum of a time series with a strong seasonal pattern will have peaks at the seasonal frequencies. If the series is adjusted by the filter in (5.15), the otherwise smooth shape of its spectrum will be punctuated by dips at these frequencies. Is this what seasonal adjustment should be doing, or are the dips in the spectrum an indication of some kind of overadjustment? There is no obvious answer to this question, but reference should be made to Nerlove *et al.* (1979, Chapter 8) and Burman (1980) for further discussion.

Spurious Cyclical Behaviour

Over half a century ago, it was observed that certain operations on a time series could produce a curious distorting effect. In particular, it was found that applying a set of summing operations could induce a spurious cycle in the data if a certain amount of differencing had already taken place. This phenomenon is known as the *Yule–Slutsky effect*. The manner in which it arises shows up quite clearly when the problem is analysed in the frequency domain.

A preliminary result is necessary. Suppose that two linear filters are applied to a series in succession. Let $W_1(\lambda)$ denote the frequency response function of the first filter, while $W_2(\lambda)$ is the corresponding function for the second. The overall effect of the combined operation is given by

$$W(\lambda) = W_2(\lambda)W_1(\lambda). \tag{5.17}$$

The corresponding transfer function is

$$|W(\lambda)|^2 = |W_2(\lambda)W_1(\lambda)|^2$$
$$= |W_2(\lambda)|^2|W_1(\lambda)|^2. \qquad (5.18)$$

Thus the effect on the spectrum of the original series is determined by the product of the transfer functions of the individual filters. A formal proof of (5.17) will not be given, as the result follows almost immediately from first principles; cf. (5.9).

Now consider the transfer function associated with the first difference operator, $1 - L$. This is given by

$$|W_1(\lambda)|^2 = 2(1 - \cos \lambda). \qquad (5.19)$$

In view of (5.18), the application of the first difference operator d times produces the transfer function

$$\prod_{i=1}^{d} |W_1(\lambda)|^2 = |W_1(\lambda)|^{2d} = 2^d(1 - \cos \lambda)^d. \qquad (5.20)$$

In a similar way, the summation filter, $x_t = x_t + x_{t+1}$, applied s times yields the transfer function

$$|W_2(\lambda)|^{2s} = 2^s(1 + \cos \lambda)^s. \qquad (5.21)$$

The overall effect of d differencing operations followed by s summations is therefore

$$|W(\lambda)|^2 = 2^{s+d}(1 - \cos \lambda)^d(1 + \cos \lambda)^s. \qquad (5.22)$$

The differencing operations attenuate the low frequencies, while the summations attenuate the high frequencies. The combined effect can be to emphasise certain of the intermediate frequencies. This shows up insofar as the transfer function has a peak. Differentiating (5.22) with respect to λ shows it to have a turning point at the frequency

$$\lambda_0 = \cos^{-1}[(s - d)/(s + d)], \qquad (5.23)$$

and taking the second derivative confirms that this is a maximum. By a suitable choice of d and s, λ_0 may be made arbitrarily close to any desired frequency, and the larger d and s are, the sharper the peak in the transfer function.

An interesting example of what is essentially the Yule–Slutsky effect is described by Fishman (1969, pp. 45–9). This is important in that it illustrates some of the pitfalls that may be encountered in analysing time series data. The study in question, by Kuznets (1961), was an investigation of the long swing hypothesis, which postulates the existence of a cycle of around 20 years in certain economic time series. Before proceeding to examine the data for evidence of a long-run cycle, Kuznets first decided to remove some of the cyclical components at higher frequencies. He therefore applied two filters. The first, a simple five year moving average of the form (5.1), was presumably aimed at attenuating the effects of the five year trade cycle. This filter has a frequency response function given by (5.11). The frequency response function of the second filter, a differencing operation of the form

$$y_t = x_{t+5} - x_{t-5},\tag{5.24}$$

is simply $2i \sin 5\lambda$; cf. (A.8). Combining these two results through (5.18) gives an overall transfer function of

$$|W(\lambda)|^2 = \left[\frac{2 \sin 5\lambda \sin (5\lambda/2)}{5 \sin (\lambda/2)}\right]^2.\tag{5.25}$$

This has a very high peak centred at a frequency corresponding to a period of 20.3 years. Kuznets concluded that the average period of the cycles he was observing was about 20 years, but as (5.25) shows, these movements could correspond to a spurious cycle induced by the two filtering operations!

6. Estimation of the Spectrum

The power spectrum of a stochastic process was defined in (1.1). An obvious estimator, given a set of T observations, is therefore

$$I(\lambda) = \frac{1}{2\pi}\left[c(0) + 2 \sum_{\tau=1}^{T-1} c(\tau)\cos \lambda\tau\right], \qquad 0 \leqslant \lambda \leqslant \pi.\tag{6.1}$$

The theoretical autocovariances in (1.1) are replaced by the sample autocovariances in (6.1). Since T is finite, the summation in (6.1) is no longer infinite. Autocovariances can only be estimated up to a lag of $T-1$, with $c(T-1)$ being a function of a single pair of observations, the first and the last.

Expression (6.1) defines the *sample spectral density*. It is closely related to the periodogram. From (2.16)

$$p_j = \frac{2}{T}\left|\sum_{t=1}^{T} (y_t - \bar{y})e^{-i\lambda_j t}\right|^2$$

$$= \frac{2}{T}\sum_{t=1}^{T}\sum_{s=1}^{T} (y_t - \bar{y})(y_s - \bar{y})e^{-i\lambda_j(t-s)}.$$

Letting $\tau = t - s$, and noting the definition of $c(\tau)$ in (2.1.11), gives

$$p_j = 2 \sum_{\tau=-(T-1)}^{T-1} c(\tau)e^{-i\lambda_j\tau}$$

$$= 2 \left[c(0) + 2 \sum_{\tau=1}^{T-1} c(\tau)\cos \lambda\tau \right]. \tag{6.2}$$

The last equality follows from the symmetry of $c(\tau)$. Comparing (6.2) with (6.1) then shows that

$$I(\lambda_j) = (1/4\pi)p_j \tag{6.3}$$

for λ_j defined by (2.6). If T is even, $I(\lambda_n) = p_n/2\pi$.

There is nothing to prevent $I(\lambda)$ from being computed for any value of λ in the range $[-\pi, \pi]$. However, it is generally convenient to consider it as being defined at the same frequencies for which the periodogram is calculated. This will be implicitly assumed throughout the ensuing discussion.

Since the sample spectral density, $I(\lambda)$, is proportional to the periodogram, it is often referred to as the periodogram in the literature. This usage will sometimes be adopted here when the distinction is unimportant.

Properties of the Sample Spectral Density

If the observations, $y_1, \ldots, y_T$, are drawn from a normally distributed white noise process, the properties of the sample spectral density may be derived very easily using the results in Section 2. If all the α and β parameters in (2.17) are zero, it follows from (2.22) and (6.3) that

$$I(\lambda_j) \sim (\sigma^2/4\pi) \cdot \chi_2^2 \tag{6.4}$$

for all λ_j as defined by (2.6).

The theoretical power spectrum of a white noise process is $f(\lambda) = \sigma^2/2\pi$ for all λ. Since the mean and variance of a χ_f^2 distribution are f and $2f$ respectively,

$$E[I(\lambda_j)] = (\sigma^2/4\pi)E(\chi_2^2) = \sigma^2/2\pi = f(\lambda) \tag{6.5}$$

while

$$\text{Var}[I(\lambda_j)] = \frac{\sigma^4}{16\pi^2} \text{Var}(\chi_2^2) = \left(\frac{\sigma^2}{2\pi}\right)^2 = f^2(\lambda). \tag{6.6}$$

For a given λ_j, the sample spectral density is an unbiased estimator of $f(\lambda)$. However, its variance does not depend on T and so

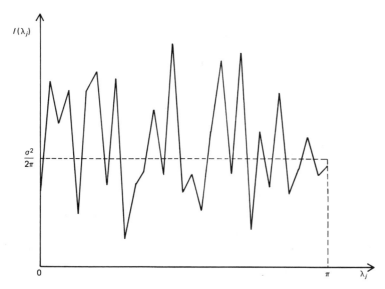

Figure 3.9   *Sample Spectral Density for a White Noise Process*

it does not give a consistent estimator of the power spectrum at a
given frequency. Furthermore, since the estimators of the parameters
in (2.17) are mutually uncorrelated, it follows that

$$\text{Cov}[I(\lambda_j), I(\lambda_i)] = 0, \qquad \text{for } j \neq i. \tag{6.7}$$

Thus with normally distributed observations, the ordinates of the
sample spectral density are independent. As a result of these
properties, a graph of the sample spectral density calculated at the
points defined by (2.6) has a jagged and irregular appearance. This is
illustrated in figure 3.9. Such a pattern persists no matter how large
the sample.

   If the observations are generated by any stationary stochastic
process with a continuous spectrum, it can be shown that $2I(\lambda)/f(\lambda)$
has a limiting χ_2^2 distribution for $0 \leqslant \lambda \leqslant \pi$. Thus the limiting
distribution of $I(\lambda)$ has a mean of $f(\lambda)$ and a variance of $f^2(\lambda)$. The
ordinates of the sample spectral density at different frequencies are
asymptotically independent. This suggests that $I(\lambda)$ will continue to
have the irregular appearance noted in the white noise model. The
empirical evidence confirms that this is indeed the case.

Spectrum Averaging

For a white noise process, the independence of the ordinates of the
sample spectral density at the frequencies defined by (2.6) suggests

smoothing $I(\lambda_j)$ by averaging over adjacent frequencies. Thus a possible estimator of $f(\lambda)$ at $\lambda = \lambda_j$ is

$$\hat{f}(\lambda_j) = m^{-1} \sum_{i=-m^*}^{m^*} I(\lambda_{j-i}) \tag{6.8}$$

where $m = 2m^* + 1$. In view of the additive properties of the χ^2 distribution it follows from (6.8) that

$$\hat{f}(\lambda_j) \sim (\sigma^2/4\pi m)\chi^2_{2m}. \tag{6.9}$$

The expectation of $\hat{f}(\lambda_j)$ is $\sigma^2/2\pi$, and so it is unbiased. Its variance is

$$\text{Var}[\hat{f}(\lambda_j)] = (\sigma^2/2\pi)^2/m = f^2(\lambda_j)/m \tag{6.10}$$

and although T does not enter directly into (6.10), it could be made to do so by allowing m to increase as T increases. To formalise this, set $m = \kappa T$, where κ is a constant. Expression (6.10) then becomes

$$\text{Var}[\hat{f}(\lambda_j)] = [f^2(\lambda_j)/\kappa]/T. \tag{6.11}$$

Since the variance of $\hat{f}(\lambda_j)$ is $0(T^{-1})$, it is a consistent estimator of $f(\lambda_j)$.

A convenient way of computing (6.8) at all the points defined by (2.6) is by a simple moving average. This yields estimates at n points, although only those estimates which are at least m points apart will be independent.

The estimator defined by (6.8) uses what is known as a rectangular *window*. The term 'window' conveys the notion that only some of the sample spectral ordinates are taken account of, or 'seen', in the weighting procedure. In this case the window is rectangular, as all the ordinates which are 'seen' are given equal weight in the construction of $\hat{f}(\lambda_j)$.

The *width* of the window is simply the length of its base. The width may be measured in two ways. The first measure is the *range*, which is the number of spectral points used in each weighted average. Thus the range is equal to m. There are n points in the spectrum, and so dividing this figure by m gives the number of completely independent estimators. The second measure is the *bandwidth*, which is the width of the window in radians. If $T = 200$ and $m = 5$, it is possible to construct 20 independent estimators of the spectrum in the range $[0, \pi]$ and so the width of each window is $\pi/20$. In general, the bandwidth is equal to $2\pi m/T$.

For a white noise spectrum, any weighted average will be an unbiased estimator of $f(\lambda)$. Furthermore, it is apparent from (6.10) that for a given sample size, the variance of the estimator will be smaller the larger is m. This inverse relationship between m and the variance of the estimator holds, irrespective of whether or not the series is generated by a white noise process. However, once we relax the assumption that the underlying spectrum is flat, it is no longer the case that an estimator like (6.8) will always be unbiased. Furthermore, a trade-off now emerges with respect to the choice of m, since the wider the bandwidth, the greater the possibility of a large bias emerging. A large value of m also leads to sharp peaks in the spectrum being 'smudged' over a much greater number of spectral points than should ideally be the case. Hence a sharp cyclical pattern may not emerge very clearly. The problems associated with too wide a bandwidth are captured by the term *resolution*. Poor resolution implies a good deal of smudging and bias.

The discussion so far has been with respect to the simple weighted average, (6.8). Other weighting schemes may also be considered, and (6.8) may be generalised to

$$\hat{f}(\lambda_j) = \sum_{i=-m^*}^{m^*} h_i I(\lambda_{j-i}), \tag{6.12}$$

where the weights, h_i, sum to unity. One possible scheme is to choose the weights in such a way that the window is triangular. This gives more weight to the ordinates closest to the frequency of interest.

The estimator given by (6.12) is an unbiased estimator of the spectrum of a white noise series, and its variance is not difficult to obtain in this case. The bandwidth for such an estimator is generally defined as being equal to the width of a rectangular window which would produce an estimator with the same variance.

Weighting in the Time Domain

An alternative approach to spectral estimation is to weight the autocovariances. The sample spectral density, (6.1), is modified by a series of weights $k(0), k(1), \ldots, k(T-1)$, to give

$$\tilde{f}(\lambda) = (2\pi)^{-1} \left[k(0)c(0) + 2 \sum_{\tau=1}^{T-1} k(\tau)c(\tau)\cos \lambda\tau \right], \qquad 0 \leqslant \lambda \leqslant \pi. \tag{6.13}$$

These weights are known as the *lag window*.

The simplest weighting scheme consists of truncating the weights beyond a certain lag length, N. For $|\tau| \leqslant N$, the weights are set equal to unity and so the lag window is effectively rectangular. This is intuitively sensible, since it cuts out the higher order sample auto-covariances and these are the ones which are most unstable.

Unfortunately, an estimator constructed in this way exhibits certain undesirable properties. This may be demonstrated by seeing what this particular weighting scheme implies in the frequency domain. Corresponding to any lag window there is a spectral window. This is exactly the same as the window described in the previous section, except that it may be regarded as a continuous function of λ. It is convenient to define λ in terms of the difference between the frequency of interest and a particular adjacent frequency, so that the expression

$$h(\lambda) = (2\pi)^{-1}\left[k(0) + 2 \sum_{\tau=1}^{T-1} k(\tau)\cos \lambda\tau\right], \qquad -\pi \leqslant \lambda \leqslant \pi$$

(6.14)

defines the *spectral window* centred on $\lambda = 0$. Since $h(\lambda)$ is symmetric, it need only be defined for positive λ, and it is almost invariably plotted on this basis. The derivation of (6.14) is carried out by replacing $c(\tau)$ in (6.13) by its Fourier transform and re-arranging the resulting expression.

For the truncated estimator

$$h(\lambda) = (2\pi)^{-1}\left[1 + 2 \sum_{\tau=1}^{N} \cos \lambda\tau\right]$$

$$= \frac{\sin[(N + 1/2)\lambda]}{2\pi \sin(\lambda/2)}.$$

(6.15)

The last equality follows from (A.9). Figure 3.10 shows the spectral window for $N = 6$. A larger value of N would result in $h(\lambda)$ being more concentrated around $\lambda = 0$, but at the expense of increasing the variance. As in the direct averaging of the sample spectral density, there is a trade-off between bias and variance. However, the fluctuating shape of the window in figure 3.10 means that the nature of the bias is rather different from that induced, say, by a rectangular window in the frequency domain. In the latter case the peaks and troughs in the spectrum tend to be smoothed out, but with the truncated estimator the existence of subsidiary peaks in the window means that an estimate may reflect important cycles in another part of the spectrum. This is known as *leakage*. When a relatively large peak corresponds to a negative part of the window, it is even possible for $\tilde{f}(\lambda)$ itself to be negative.

By suitably modifying the weighting scheme, the problems inherent in the truncated estimator can, to a large extent, be overcome. A number of estimation procedures have been suggested, two of the most widely used being due to Blackman and Tukey, and Parzen.

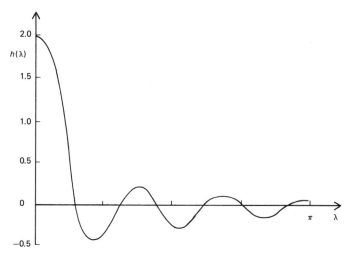

Figure 3.10 *Spectral Window, (6.15), for the Truncated Estimator with N = 6*

Both truncate the weights beyond some pre-specified lag, N, but the values of $k(\tau)$ decline as τ moves from zero to N. Less emphasis is given to the higher order autocovariances and this tends to produce an estimator with better resolution.

Fast Fourier Transform

Direct averaging of the sample spectral density is conceptually the most straightforward way of computing a consistent estimator of the spectrum. Unfortunately, the task of computing all n periodogram ordinates is, at first sight, a relatively time-consuming one, and historically this led to a shift in emphasis towards weighting the sample autocovariance function in the time domain. Since the mid-sixties, however, a technique known as the *Fast Fourier Transform* has become increasingly popular. This reduces the computing time for Fourier coefficients from $0(T^2)$ to $0(T\log_2 T)$, a reduction which can be considerable for a moderately large value of T. The availability of the Fast Fourier Transform algorithm has tended to re-establish periodogram averaging, although time domain weighting is still widely used.

Autoregressive Spectral Estimation

Any linear process can be approximated by an AR process, and this has led Parzen (1969) to suggest a technique known as *autoregressive spectral estimation*. The first step in this procedure is to estimate the coefficients of an AR(P) process by OLS. These estimates are then substituted for the parameters, $\phi_1, \phi_2, \ldots, \phi_P$ in expression (4.12), thereby yielding an estimator of the power spectrum, $f(\lambda)$. In other words, an AR(P) model is fitted to the data, and the estimator of the power spectrum is taken as the *theoretical* spectrum of the fitted process.

Just as the lag length, N, is allowed to increase with T when estimation is based on the weighted autocovariance function, so P may be allowed to increase with T, and for large T, Parzen shows that

$$\text{Var}[\,\hat{f}(\lambda)] \simeq [2Pf^2(\lambda)]/T. \tag{6.16}$$

The fact that the variance is proportional to P suggests that P be made no larger than is necessary for the AR model to be an adequate approximation to the underlying data generation process. The difficulty lies in deciding on a value for P, since if it is made too small the estimated spectrum may be badly biased, with the result that the salient frequency domain characteristics of the series will be seriously distorted. The trade-off is therefore similar to that encountered in other types of spectral estimation. One solution is to actively determine the order of the model on a goodness of fit criterion, rather than employ an automatic rule, which, say, sets $P = \sqrt{T}$. Criteria which have been adopted for determining the value of P include the AIC, defined in (1.3.2); cf. Akaike (1969; 1974). Such methods have been used in autoregressive spectral estimation with some success. Because the estimator is a theoretical spectrum, it will tend to be smoother than a spectrum produced by standard methods. However, although the procedure gives a smooth spectrum, it also seems to have high resolution, in that it is able to pick out narrow peaks; see, for example, Griffiths and Prieto-Diaz (1977). This is important, since the main *raison d'être* of spectral analysis is in the description it provides of the cyclical movements in a series. Nevertheless, great care must be taken in determining the value of P, since setting it too low could result in a failure of the spectrum to show the narrow peaks indicative of strong cyclical movements.

7. Multivariate Spectral Analysis

The time domain properties of a stationary vector process were examined in Section 2.6. In the frequency domain, the analogue of the $N \times N$ autocovariance matrix is the $N \times N$ *multivariate spectrum*

$$F(\lambda) = (2\pi)^{-1} \sum_{\tau=-\infty}^{\infty} \Gamma(\tau)e^{-i\lambda\tau}, \qquad -\pi \leqslant \lambda \leqslant \pi. \tag{7.1}$$

The diagonal elements of $F(\lambda)$ are the power spectra of the individual processes. The ijth element of $F(\lambda)$ is the *cross-spectrum* between the ith and the jth variable for $j > i$. It is the cross-spectrum which contains all the information concerning the relationship between two series in the frequency domain. The jith element of $F(\lambda)$ is simply the complex conjugate of the ijth element.

Gain, Phase and Coherence

The relationship between two series is normally characterised by the gain and the phase. These two quantities are derived from the cross-spectrum, but they are real rather than complex. Suppose that y_t and x_t are jointly stationary stochastic processes, with continuous power spectra $f_y(\lambda)$ and $f_x(\lambda)$ respectively. The cross-spectrum between y_t and x_t is defined by

$$f_{yx}(\lambda) = (2\pi)^{-1} \sum_{\tau=-\infty}^{\infty} \gamma_{yx}(\tau)e^{-i\lambda\tau} \tag{7.2}$$

for $-\pi \leqslant \lambda \leqslant \pi$. The *gain* is

$$G(\lambda) = |f_{yx}(\lambda)|/f_x(\lambda), \tag{7.3}$$

while the *phase* is

$$\text{Ph}(\lambda) = \tan^{-1}\{-Im[f_{yx}(\lambda)]/Re[f_{yx}(\lambda)]\}. \tag{7.4}$$

The concepts of gain and phase were introduced in Section 5 to describe the properties of a linear filter. Their interpretation in cross-spectral analysis is similar. Suppose that (4.1) is modified by adding to the right-hand side a stochastic disturbance term which is distributed independently of x_t. This yields the model

$$y_t = \sum_{j=-r}^{s} w_j x_{t-j} + u_t, \qquad t = 1, \ldots, T. \tag{7.5}$$

If, for convenience, x_t is assumed to have zero mean, multiplying both sides of (7.5) by $x_{t-\tau}$ and taking expectations gives

$$\gamma_{yx}(\tau) = \sum_{j=-s}^{r} w_j \gamma_x(\tau - j), \qquad \tau = 0, \pm 1, \pm 2, \ldots \tag{7.6}$$

Substituting (7.6) into (7.2) gives

$$f_{yx}(\lambda) = (2\pi)^{-1} \sum_{\tau=-\infty}^{\infty} \sum_{j=-r}^{s} w_j \gamma_x(\tau - j)e^{-i\lambda\tau}.$$

Setting $k = \tau - j$ and reversing the order of the summation signs then yields

$$f_{yx}(\lambda) = (2\pi)^{-1} \sum_{j=-r}^{s} w_j e^{-i\lambda j} \sum_{k=-\infty}^{\infty} \gamma_x(k)e^{-i\lambda k}$$

which on comparing with (4.2) is seen to be

$$f_{yx}(\lambda) = W(\lambda)f_x(\lambda). \tag{7.7}$$

Having established (7.7), the link between (7.3) and (7.4) and the formulae given in Section 5 is clear. The gain and phase characterise the relationship between y and x, just as they do for the linear filter, (4.1). The only difference is that the relationship is no longer an exact one. This suggests the need for a third quantity which somehow measures the strength of the relationship between y_t and x_t at different frequencies.

Let z_t denote the systematic part of y_t, i.e.

$$z_t = \sum_{j=-r}^{s} w_j x_{t-j}, \qquad t = 1, \ldots, T. \tag{7.8}$$

Since $E(x_t u_\tau) = 0$ for all t and τ

$$\gamma_y(\tau) = \gamma_z(\tau) + \gamma_u(\tau), \qquad \tau = 0, 1, 2, \ldots \tag{7.9}$$

and so

$$f_y(\lambda) = f_z(\lambda) + f_u(\lambda). \tag{7.10}$$

The *coherence*,

$$\text{Coh}(\lambda) = f_z(\lambda)/f_y(\lambda) = 1 - f_u(\lambda)/f_y(\lambda), \qquad 0 \leqslant \lambda \leqslant \pi, \tag{7.11}$$

is a measure of the fraction of $f_y(\lambda)$ which can be systematically accounted for by movements in x. Since

$$f_z(\lambda) = |W(\lambda)|^2 f_x(\lambda) \tag{7.12}$$

(cf. (4.3)), the coherence can be expressed in an alternative form by substituting from (7.7), i.e.

$$\text{Coh}(\lambda) = \frac{|f_{yx}(\lambda)|^2}{f_x(\lambda)f_y(\lambda)}. \tag{7.13}$$

Example 1   Consider the model defined by (2.6.6) and (2.6.8).

Since

$$W(\lambda) = \beta e^{-i\lambda}, \tag{7.14}$$

it follows immediately from (5.8) that $G(\lambda) = \beta$ at all frequencies while $\mathrm{Ph}(\lambda) = \lambda$. After some re-arrangement, the coherence function may be written as

$$\mathrm{Coh}(\lambda) = \left[1 + \frac{f_u(\lambda)}{|W(\lambda)|^2 f_x(\lambda)} \right]^{-1}, \tag{7.15}$$

and so in this case

$$\mathrm{Coh}(\lambda) = [1 + \sigma^2 \beta^{-2} \sigma_\eta^{-2} (1 + \phi^2 - 2\phi \cos \lambda)]^{-1}. \tag{7.16}$$

This is shown in figure 3.11 for $\phi = 0.8$, $\beta = 1$ and $\sigma^2 = \sigma_\eta^2 = 1$. Since x_t is slowly changing, its spectrum has relatively high power at the lower frequencies and this is reflected in the high coherence around $\lambda = 0$. On the other hand, the disturbance term has a much greater effect at the higher frequencies, and the relationship between x and y is correspondingly weaker. Note that ϕ is the crucial parameter in determining the overall shape of the coherence function, and that a positive value is likely to be more characteristic of actual time series than a negative one. The pattern shown in figure 3.11 is probably not atypical.

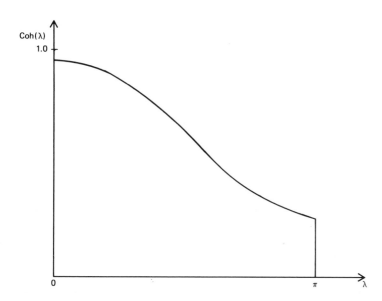

Figure 3.11   *Coherence between x_t and y_t in Example 7.1*

Example 2   The Koyck, or geometric, distributed lag model,

$$y_t = \beta \sum_{j=0}^{\infty} \alpha^j x_{t-j} + \epsilon_t, \qquad 0 < \alpha < 1, \tag{7.17}$$

imposes an exponentially declining pattern on the lag coefficients. In terms of (7.5), $w_j = \beta\alpha^j$, and the frequency response function is given by

$$W(\lambda) = \beta \sum_{j=0}^{\infty} \alpha^j e^{-i\lambda j} = \beta/(1 - \alpha e^{-i\lambda}). \tag{7.18}$$

If x_t is generated by the AR(1) process in (2.6.8), the coherence is given by

$$\text{Coh}(\lambda) = [1 + \sigma^2\beta^{-2}\sigma_\eta^{-2}(1 + \phi^2 - 2\phi \cos \lambda)(1 + \alpha^2 - 2\alpha \cos \lambda)]. \tag{7.19}$$

When ϕ is positive, the salient characteristic of figure 3.11, namely the relatively high coherence at low frequencies, is reproduced by (7.19). This feature becomes more exaggerated the closer α is to one.

Estimation

Given data on two time series, the gain, phase and coherence functions may all be estimated. As they stand, they provide a description of the relationship between the two series in the frequency domain. There need be no assumption that the series are linked by the kind of linear relationships which have featured in the discussions so far; these were introduced merely to illustrate certain aspects of the theoretical cross-spectrum. Indeed, one of the attractions of cross-spectral analysis is that it may permit the characterisation of cyclical relationships which are difficult to model in the time domain.

Since the cross-spectrum is a complex quantity, it is normally broken down into two real quantities, the *co-spectrum*, $c(\lambda)$, and the *quadrature spectrum*, $q(\lambda)$. Thus

$$f_{yx}(\lambda) = c(\lambda) + iq(\lambda), \tag{7.20}$$

where

$$c(\lambda) = (2\pi)^{-1} \sum_{\tau=-\infty}^{\infty} \gamma_{yx}(\tau)\cos \lambda\tau \tag{7.21a}$$

and

$$q(\lambda) = -(2\pi)^{-1} \sum_{\tau=-\infty}^{\infty} \gamma_{yx}(\tau)\sin \lambda\tau. \qquad (7.21\mathrm{b})$$

A direct interpretation of $c(\lambda)$ and $q(\lambda)$ will be found in Granger and Hatanaka (1964, pp. 74–6). However, the information in the cross-spectrum is more easily analysed in terms of gain, phase and coherence and these quantities can be readily obtained from $c(\lambda)$ and $q(\lambda)$.

The main considerations which arise in estimating the cross-spectrum are essentially the same as those which arise with the spectrum. The sample co-spectrum and quadrature spectrum are obtained by replacing the theoretical autocovariances in (7.21) by the corresponding sample autocovariances. However, consistent estimators of the gain, phase and coherence can only be obtained if some smoothing is carried out. This may be done directly in the frequency domain by averaging over adjacent frequencies, or in the time domain by using an appropriate lag window to weight the sample cross-covariances.

The sampling distributions of the gain, phase and coherence are discussed at some length in Fishman (1969, pp. 131–8). An important point concerns the dependence of the large sample variances of these statistics on the population coherence. The closer $\mathrm{Coh}(\lambda)$ is to unity at a particular frequency, the smaller the variance of the corresponding sample statistics. On the other hand when the coherence is small, the estimated gain, phase and coherence may all be unreliable. This may not be surprising once it is realised that the sample coherence can be interpreted as a measure of the 'correlation' between the two series at different frequencies.

Appendix A   Trigonometric Identities

The *addition formulae* are

$$\sin(x \pm y) = \sin x \cos y \pm \cos x \sin y \qquad (\mathrm{A.1})$$

$$\cos(x \pm y) = \cos x \cos y \mp \sin x \sin y. \qquad (\mathrm{A.2})$$

The addition formulae form the basis for deriving many results. For example, setting $x = y$ in (A.2) and using the fundamental identity

$$\sin^2 x + \cos^2 x = 1,$$

yields, after some re-arrangement,

$$\cos^2 x = \tfrac{1}{2} + \tfrac{1}{2} \cos 2x. \tag{A.3}$$

Such results are often useful in the calculus. For example

$$\int \cos^2 x \, dx = \int \left(\tfrac{1}{2} + \tfrac{1}{2} \cos 2x\right) dx$$

$$= \tfrac{1}{2}x + \tfrac{1}{4} \sin 2x = \tfrac{1}{2}(x + \sin x \cos x). \tag{A.4}$$

The identity

$$e^{ix} = \cos x + i \sin x \tag{A.5}$$

plays a particularly important role in spectral analysis, one of the main reasons being that the exponential function is relatively easy to manipulate. Thus, given (A.5), the first addition formula may be shown to hold directly by expanding both sides of the equation

$$e^{i(x+y)} = e^{ix} e^{iy} \tag{A.6}$$

and equating real and complex parts.

Because of the convenience of working with exponential terms, the sine and cosine function are often expressed in the form

$$\cos x = (e^{ix} + e^{-ix})/2, \tag{A.7}$$

and

$$\sin x = (e^{ix} - e^{-ix})/2i. \tag{A.8}$$

Both results follow almost directly from (A.5).

An example of the use of these formulae is given by the proof of the identity

$$y_n(\lambda) = \tfrac{1}{2} + \sum_{x=1}^{n} \cos \lambda x = \frac{\sin(n + 1/2)\lambda}{2 \sin(\lambda/2)}. \tag{A.9}$$

The first step is to use (A.7) to construct

$$y_n(\lambda) = \frac{1}{2} \sum_{x=-n}^{n} e^{i\lambda x} = \tfrac{1}{2} e^{-in\lambda}[1 + e^{i\lambda} + e^{i2\lambda} + \cdots + e^{i2n\lambda}]. \tag{A.10}$$

The term in square brackets may be summed as a finite geometric progression. This yields

$$y_n(\lambda) = \tfrac{1}{2} e^{-in\lambda} \frac{1 - e^{i\lambda(2n+1)}}{1 - e^{i\lambda}} = \frac{1}{2} \frac{e^{-in\lambda} - e^{i(n+1)\lambda}}{1 - e^{i\lambda}}.$$

Multiplying the numerator and denominator of this expression by $-e^{i\lambda/2}$ and using (A.8) gives the required result.

Appendix B   Orthogonality Relationships

The trigonometric orthogonality relationships between pairs of sines and cosines may be expressed concisely in the form

$$\int_{-\pi}^{\pi} e^{inx} e^{-imx} \, dx = \begin{cases} 0, & n \neq m \\ 2\pi, & n = m \end{cases} \tag{B.1}$$

where n and m are integers. When $n = m$ this result follows directly, while for $n - m = k \neq 0$,

$$\int_{-\pi}^{\pi} e^{ikx} = \left[\frac{1}{ik} e^{ikx} \right]_{-\pi}^{\pi} = 0 \tag{B.2}$$

since

$$e^{ik\pi} = e^{-ik\pi} = \begin{cases} 1 & \text{for } k \text{ even} \\ -1 & \text{for } k \text{ odd.} \end{cases}$$

The orthogonality relationships used in deriving the Fourier representation of Section 2 are:

$$\sum_{t=1}^{T} \cos \frac{2\pi j}{T} t \cdot \cos \frac{2\pi k}{T} t = \begin{cases} 0, & 0 \leqslant k \neq j \leqslant [T/2] \\ T/2, & 0 < k = j < T/2 \\ T, & k = j = 0, T/2 \end{cases} \tag{B.3a}$$

$$\sum_{t=1}^{T} \sin \frac{2\pi j}{T} t \cdot \sin \frac{2\pi k}{T} t = \begin{cases} 0, & 0 \leqslant k = j \leqslant [T/2] \\ T/2, & 0 < k \neq j < T/2 \\ 0, & k = j = 0, T/2 \end{cases} \tag{B.3b}$$

$$\sum_{t=1}^{T} \cos \frac{2\pi j}{T} t \cdot \sin \frac{2\pi k}{T} t = 0, \quad k, j = 0, 1, \ldots, [T/2], \tag{B.3c}$$

where $[T/2] = T/2$ for T even and $(T-1)/2$ for T odd.
Setting $j = 0$ in (B.3a) and (B.3c) yields

$$\sum_{t=1}^{T} \cos \frac{2\pi k}{T} t = 0, \quad k = 1, \ldots, [T/2] \tag{B.4a}$$

and

$$\sum_{t=1}^{T} \sin \frac{2\pi k}{T} t = 0, \quad k = 0, 1, \ldots, [T/2] \tag{B.4b}$$

respectively. Using these results together with (A.5) shows that

$$\sum_{t=1}^{T} e^{i\lambda t} = 0. \tag{B.5}$$

Appendix C   Fourier Transforms

The results in (B.1) and (B.2) form the basis for the *Fourier transform* of a continuous function. The expression

$$f(x) = \frac{\alpha_0}{2} + \sum_{k=1}^{n} (\alpha_k \cos kx + \beta_k \sin kx) \tag{C.1}$$

is a trigonometric polynomial of order n, with $2n + 1$ Fourier coefficients, $\alpha_0, \alpha_1, \ldots, \alpha_n, \beta_1, \ldots, \beta_n$. Since $f(x)$ is a linear combination of sines and cosines of period 2π, it follows that it is a continuous function over any range of x of length 2π. By defining

$$\delta_k = (\alpha_k + i\beta_k)/2, \quad \delta_{-k} = (\alpha_k - i\beta_k)/2 \quad \text{and} \quad \delta_0 = \alpha_0/2,$$

expression (C.1) may be written in complex notation as

$$f(x) = \sum_{k=-n}^{n} \delta_k e^{-ikx}. \tag{C.2}$$

Suppose now that $f(x)$ is given, and that we wish to find the δ_k's or, equivalently, the α_k's and β_k's. Multiplying both sides of (C.2) by e^{ijx} and integrating over a range of 2π, say $-\pi$ to π, yields

$$\int_{-\pi}^{\pi} f(x)e^{ijx}\, dx = \sum_{k=1}^{n} \delta_k \int_{-\pi}^{\pi} e^{-i(k-j)x}\, dx, \qquad j = 0, 1, \ldots, n. \tag{C.3}$$

Using (B.1) then gives

$$\delta_k = \frac{1}{2\pi} \int_{-\pi}^{\pi} f(x)e^{ikx}\, dx. \tag{C.4}$$

Expressions (C.2) and (C.4) are often referred to as Fourier transform pairs.

If $f(x)$ is an even function, (C.2) contains only cosine terms and δ_k is real with $\delta_k = \delta_{-k}$. The Fourier transform pair may then be written:

$$f(x) = \delta_0 + 2 \sum_{k=1}^{n} \delta_k \cos kx \tag{C.5}$$

and

$$\delta_k = \frac{\alpha_k}{2} = \frac{1}{\pi} \int_{0}^{\pi} f(x)\cos kx\, dx. \tag{C.6}$$

It is always possible to approximate a continuous function by a trigonometric polynomial. As an example consider the function

$$f(x) = \begin{cases} 1/2b, & |x| \leqslant b \\ 0, & \text{elsewhere.} \end{cases} \tag{C.7}$$

Applying (C.6) yields

$$2\delta_k = \alpha_k = (2\pi b)^{-1} \int_{-b}^{b} \cos kx \, dx = \frac{\sin bk}{\pi bk} \tag{C.8}$$

and $\alpha_0 = 1/\pi$. Therefore

$$f(x) = \frac{1}{2\pi} + \frac{1}{b\pi} \sum_{k=1}^{\infty} \frac{\sin bk}{k} \cos kx. \tag{C.9}$$

Notes

Fishman (1969) and Granger and Hatanaka (1964) are good introductions to spectral analysis. Fuller (1976), Anderson (1971) and Hannan (1970) contain proofs of various key theorems.

Exercises

1. Derive the spectrum for an MA(2) process. Can this give rise to cyclical behaviour of the form exhibited by the AR(2) process?

2. Find the spectrum for $y_t = \phi y_{t-12} + \epsilon_t$, where $|\phi| < 1$, and comment on its shape.

3. Suppose that the filter in (5.1) is applied and the filtered series is then subtracted from the original series. Find the gain and compare this with the gain of (5.3).

4. Write down an expression for the phase, $\text{Ph}(\lambda)$, in terms of the co-spectrum and quadrature spectrum.

5. Derive (5.16) and (5.23).

6. Draw the power spectra for AR(1) processes with $\phi = 0.5$, $\phi = 0.99$ and $\phi = -0.5$. Comment.

7. Derive an expression for the power spectrum of an ARMA(1, 1) process. Sketch the spectra of the processes whose autocorrelation functions are shown in figure 2.5.

8. (a) Explain carefully what is meant by the 'gain' and 'phase' of a linear filter.

(b) By calculating the appropriate gain and phase functions, explain the advantages and disadvantages of using a difference filter of the form

$$y_t^* = y_t - y_{t-7},$$

as opposed to a simple seven year moving average, to remove a seven year cycle.
[University of Kent, 1976.]

9. (a) What criteria are used to evaluate different estimators of the spectral density? In what way, or ways, is the Blackman—Tukey estimator superior to the 'truncated' estimator?

(b) Derive an expression relating weighting in the time domain to weighting in the frequency domain. Hence show that the spectral window given by the truncated estimator is

$$K(\lambda) = \frac{\sin(N + 1/2)\lambda}{2\pi \sin(\lambda/2)}$$

where N denotes the point of truncation (i.e. the point beyond which all auto-covariances are given a weight of zero). [University of Kent, 1976.]

10. Determine the forms of the autocorrelation function and power spectrum of the stochastic process which results when first differences are taken in the model,

$$y_t = \alpha + \beta t + u_t,$$

$$u_t = \phi u_{t-4} + \epsilon_t,$$

where α, β and ϕ are parameters and $|\phi| < 1$.

11. The following figures are quarterly observations on an economic time series:

Year	Q.1	Q.2	Q.3	Q.4
1	2	5	9	3
2	1	3	6	5
3	3	4	8	6
4	4	4	7	6

Estimate the seasonal pattern in this series by fitting an appropriate number of trigonometric terms.

If the model is of the form (2.17), test whether the individual trigonometric coefficients are significantly different from zero, *given* that the variance of the disturbance term, ϵ_t, is known to be 2.25. Comment on the implications of your result. How would you modify the above testing procedure if σ^2 were unknown?

4
State Space Models and the Kalman Filter

1. State Space

State space models were originally developed by control engineers. In a typical application, attention is focused on a set of m *state variables* which change over time. These variables may be a signal, denoting, for example, the position of a rocket. In most cases the signal will not be directly observable, being subject to systematic distortion as well as contamination by 'noise'.

The N variables which actually are observed are defined by an $N \times 1$ vector, y_t, and these are related to the state variables by a *measurement equation*. If the state variables themselves are contained within an $m \times 1$ vector, α_t, this equation may be written as

$$y_t = Z_t \alpha_t + S_t \xi_t, \qquad t = 1, \ldots, T, \qquad (1.1)$$

where Z_t and S_t are fixed matrices of order $N \times m$ and $N \times n$ respectively. The $n \times 1$ vector of disturbances, ξ_t, has zero mean and covariance matrix, H_t.

Although the state vector, α_t, is not directly observable, its movements are assumed to be governed by a well-defined process. This process is defined by the *transition equation*,

$$\alpha_t = T_t \alpha_{t-1} + R_t \eta_t, \qquad t = 1, \ldots, T, \qquad (1.2)$$

where T_t and R_t are fixed matrices of order $m \times m$ and $m \times g$ respectively, and η_t is a $g \times 1$ vector of disturbances, with mean zero and covariance matrix Q_t.

The disturbances in both the measurement and transition equations are taken to be serially uncorrelated. Furthermore, they are uncorrelated with each other for all time periods, and with the initial state vector, α_0. These assumptions may be summarised by writing

$$\begin{pmatrix} \xi_t \\ \eta_t \end{pmatrix} \sim \mathrm{WN} \left[0, \begin{pmatrix} H_t & 0 \\ 0 & Q_t \end{pmatrix} \right], \qquad t = 1, \dots, T \tag{1.3}$$

and

$$E[\alpha_0 \eta_t'] = 0, \quad E(\alpha_0 \xi_t') = 0, \qquad t = 1, \dots, T, \tag{1.4}$$

where WN stands for 'white noise'; cf. (2.7.2).

The above representation of a linear dynamic model is known as the *state space form*. At first sight it may appear to have little connection with any of the models considered in earlier chapters. However, in the last part of this section it is shown that any ARMA model may be cast in state space form. This has important implications for ML estimation, as well as providing a framework for the treatment of time series models with unobservable components. A connection also exists between regression models and the state space form, and this may be exploited in modelling systems with time-varying parameters.

The Kalman Filter

The Kalman filter is a set of equations which allows an estimator to be updated once a new observation becomes available. This process is carried out in two parts. The first step consists of forming the optimal predictor of the next observation, given all the information currently available. This is effected by means of the *prediction* equations. The new observation is then incorporated into the estimator of the state vector using the *updating* equations.

The Kalman filter provides an *optimal* solution to the problems of prediction and updating. If the observations are normally distributed, and the current estimator of the state vector is the best available, the predictor and the updated estimator will also be the best available. In the absence of the normality assumption a similar result holds, but only within the class of estimators and predictors which are linear in the observations. Since the state vector is stochastic, some care must be taken over the interpretation of 'best' in this context. Attention must be focused on the estimation error, rather than on the estimator itself. This leads to the concept of the *minimum mean square (linear) estimator*, which is defined formally in the next section.

The filtering equations may be applied recursively as each new observation becomes available. As a by-product, a series of prediction errors are produced. The implications of this for ML estimation should be immediately apparent. In Section 1.2 it was shown that the likelihood function for a set of dependent observations may be decomposed in terms of the prediction errors. This decomposition is fundamental, and the Kalman filter provides a natural mechanism for carrying it out.

When all the observations have been processed, recursive techniques may be applied in reverse to solve the problem of *smoothing*. The updating equations give the best estimators of the state variables based on the information available at that time. The smoothed estimators, on the other hand, use all the information. Smoothing therefore provides the optimal means of extracting estimates of the state variables from the observations.

State Space Representation of an ARMA Model

The general ARMA(p, q) model, (2.1.25), can be written in the form

$$y_t = \phi_1 y_{t-1} + \cdots + \phi_m y_{t-m} + \epsilon_t + \theta_1 \epsilon_{t-1} + \cdots + \theta_{m-1} \epsilon_{t-m+1},$$

$$(1.5)$$

where $m = \max(p, q + 1)$. A Markovian representation of (1.5) is then obtained by defining an $m \times 1$ vector, α_t, which obeys the multivariate AR(1) model

$$\alpha_t = \begin{bmatrix} \phi_1 & & \\ \phi_2 & & \\ \vdots & & I_{m-1} \\ \cdots & \cdots & \cdots \\ \phi_m & & 0' \end{bmatrix} \alpha_{t-1} + \begin{bmatrix} 1 \\ \theta_1 \\ \vdots \\ \\ \theta_{m-1} \end{bmatrix} \epsilon_t. \qquad (1.6)$$

This may be regarded as a transition equation, (1.2), in which T_t and R_t are constant and $Q_t = \sigma^2$. The original ARMA model may easily be recovered from (1.6) by noting that the first element of α_t is identically equal to y_t. This can be seen by repeated substitution, starting at the bottom row of the system.

Example 1   The Markovian representation of the MA(1) model is

$$\alpha_t = \begin{bmatrix} 0 & 1 \\ 0 & 0 \end{bmatrix} \alpha_{t-1} + \begin{bmatrix} 1 \\ \theta \end{bmatrix} \epsilon_t. \qquad (1.7)$$

If $\alpha_t = (\alpha_{1t} \alpha_{2t})'$, then $\alpha_{2t} = \theta \epsilon_t$ and $\alpha_{1t} = \alpha_{2,t-1} + \epsilon_t = \epsilon_t + \theta \epsilon_{t-1}$. This is precisely the original model.

All that is required of the measurement equation is to extract the first element of the state vector. Thus (1.1) is a single equation in which the disturbance term, ξ_t, is identically equal to zero, i.e.

$$y_t = z_t'\alpha_t, \qquad t = 1, \ldots, T, \tag{1.8}$$

where $z_t' = (1 \ 0_{m-1}')'$.

A similar representation may also be given to an ARMA model obscured by measurement error. Suppose that the observations, y_t, are defined by

$$y_t = w_t + \xi_t, \qquad t = 1, \ldots, T, \tag{1.9}$$

where $w_t \sim \text{ARMA}(p, q)$ and $\xi_t \sim \text{WN}(0, \sigma^2 h)$. The transition equation is exactly the same as (1.6), but the first element in α_t is now equal to w_t and equation (1.8) includes a measurement error, ξ_t.

2. Minimum Mean Square Estimation

Consider the model

$$y = Z\alpha + \xi, \tag{2.1}$$

where y is a $T \times 1$ vector of observations, Z is a $T \times m$ matrix of fixed values and ξ is a $T \times 1$ vector of disturbances with mean zero and covariance matrix $\sigma^2\Omega$. The matrix Ω is p.d. and known. The $m \times 1$ vector, α, may be regarded as a vector of unknown parameters. As such, (2.1) has the form of a generalised regression model. However, there is an important difference in that α will be taken to be stochastic, in the sense that it is randomly drawn from a prior distribution before the observations in y are generated.

If α were fixed, the GLS estimator

$$\tilde{a} = (Z'\Omega^{-1}Z)^{-1}Z'\Omega^{-1}y, \tag{2.2}$$

would be the BLUE of α. With α random, $\tilde{a}$ again has certain optimal statistical properties, but these must now be expressed in terms of the estimation error, $\tilde{a} - \alpha$. However, the properties of $\tilde{a} - \alpha$ exactly parallel those of a conventional GLS estimator since

$$(\tilde{a} - \alpha) \sim \text{WS}[0, \sigma^2(Z'\Omega^{-1}Z)^{-1}]. \tag{2.3}$$

The notation $\text{WS}(\mu, V)$ indicates that the variable in question has mean μ and variance V; WS stands for 'wide sense'. The Gauss–Markov theorem may be modified to show that $\tilde{a}$ is 'best' within the class of estimators which are linear and unconditionally unbiased. An estimator is *unconditionally unbiased (u-unbiased)* if its estimation error has zero expectation.

Suppose, for simplicity, that $\Omega = I$. (The theorem can be proved in the more general case since there exists a nonsingular $T \times T$ matrix, with the property $L'L = \Omega^{-1}$, and this can be used to transform the observations.) If $\hat{a} = D^*y$ is any other linear estimator of α, then

$$\hat{a} - \alpha = DZ\alpha + [D + (Z'Z)^{-1}Z']\xi,$$

where $D = D^* - (Z'Z)^{-1}Z'$. For $E(\hat{a} - \alpha) = 0$, it is necessary that $DZ = 0$. With this condition imposed, the covariance matrix of $\hat{a} - \alpha$ is

$$\begin{aligned}
\text{Var}(\hat{a} - \alpha) &= E[(\hat{a} - \alpha)(\hat{a} - \alpha)'] \\
&= [D + (Z'Z)^{-1}Z']E(\xi\xi')[D' + Z(Z'Z)^{-1}].
\end{aligned}$$

Since $E(\xi\xi') = \sigma^2 I$ and $DZ = 0$, this expression simplifies to

$$\text{Var}(\hat{a} - \alpha) = \sigma^2 D'D + \sigma^2(Z'Z)^{-1}.$$

The covariance matrix of $\hat{a} - \alpha$ therefore exceeds that of $\tilde{a} - \alpha$ by a p.s.d. matrix.

When an estimator is u-unbiased its MSE matrix, $E[(\hat{a} - \alpha)(\hat{a} - \alpha)']$ is identical to the covariance matrix of the estimation error. However, if the initial criterion adopted for assessing the desirability of an estimator is that it should minimise the MSE of each element in $\hat{a} - \alpha$, the estimator must necessarily be u-unbiased. This follows on splitting the MSE into two parts:

$$E[(\hat{a}_i - \alpha_i)^2] = [E(\hat{a}_i - \alpha_i)]^2 + \text{Var}(\hat{a}_i - \alpha_i), \qquad i = 1, \dots, m.$$
(2.4)

Since the variance of $\hat{a}_i - \alpha_i$ does not depend on whether or not $\hat{a}_i$ is u-unbiased, it follows that (2.4) may be minimised in two steps. The first of these sets $[E(\hat{a}_i - \alpha_i)]^2$ equal to zero by constraining the estimator to be u-unbiased. The second determines that $\hat{a}$ is equal to $\tilde{a}$ by drawing on the above modification of the Gauss–Markov theorem. The end result, namely that $\tilde{a}$ is the *minimum mean square linear estimator* (MMSLE) of α, is known as the *extended Gauss–Markov theorem*.

If the observations are normally distributed, the optimal properties of $\tilde{a}$ are no longer confined to the class of linear estimators. In these circumstances, $\tilde{a}$ is the *minimum mean square estimator* (MMSE) of α.

The terms MMSLE and MMSE can obviously be used to describe predictors, since, in general, predictors are simply estimators of future values of random variables.

Combining Prior and Sample Information

Suppose that information on α is already available, and that this takes the form

$$a_0 - \alpha \sim WS(0, \sigma^2 P_0),\tag{2.5}$$

where a_0 is a known vector and P_0 is a known, p.d. matrix. Expressing the prior information in terms of a distribution for $a_0 - \alpha$ allows a rather general formulation, in which the elements of α may be regarded as fixed or random. If a particular element of α is fixed, the corresponding element of a_0 must be random. The problem is to incorporate the information in (2.5) with that in the sample, (2.1). The first step is to construct an augmented model

$$\begin{pmatrix} a_0 \\ y \end{pmatrix} = \begin{pmatrix} I \\ Z \end{pmatrix} \alpha + \begin{pmatrix} a_0 - \alpha \\ \xi \end{pmatrix}.\tag{2.6}$$

This may be written more concisely as

$$y^\dagger = Z^\dagger \alpha + \xi^\dagger,\tag{2.7}$$

where $E(\xi^\dagger) = 0$ and

$$E(\xi^\dagger \xi^{\dagger\prime}) = \sigma^2 V = \sigma^2 \begin{pmatrix} P_0 & 0 \\ 0 & \Omega \end{pmatrix}.\tag{2.8}$$

Although α has been re-defined to include the possibility that some, or all, of its elements are fixed, the extended Gauss–Markov theorem still applies. The estimator

$$\tilde{a}^\dagger = (Z^{\dagger\prime} V^{-1} Z^\dagger)^{-1} Z^{\dagger\prime} V^{-1} y^\dagger\tag{2.9}$$

is therefore the MMSLE of α. Using the original notation this may be re-written as

$$\tilde{a}^\dagger = P(P_0^{-1} a_0 + Z' \Omega^{-1} y),\tag{2.10}$$

where

$$P = (P_0^{-1} + Z' \Omega^{-1} Z)^{-1}.\tag{2.11}$$

Thus

$$\tilde{a}^\dagger - \alpha \sim WS(0, \sigma^2 P).\tag{2.12}$$

If all the elements of α are fixed, (2.1) is a standard generalised regression model. The incorporation of the prior information in (2.5) is then a special case of the *mixed estimation procedure* introduced by Theil and Goldberger (1961), and described in Theil (1970, pp. 346–52). On the other hand, if α is stochastic, the procedure may be given a *Bayesian* interpretation. Expression (2.5) refers to prior distribution of α, while (2.12) is the posterior distribution. Both distributions, however, are conditional on the unknown parameter σ^2.

3. Updating and Prediction Equations

In many cases of interest only one observation is available in each time period. Expression (1.1) therefore reduces to a single measurement equation,

$$y_t = z_t'\alpha_t + \xi_t, \qquad t = 1, \ldots, T, \tag{3.1}$$

where $\xi_t \sim WN(0, \sigma^2 h_t)$. The transition equation is often much simpler than (1.2), as T_t, R_t and Q_t will typically be time-invariant. Thus

$$\alpha_t = T\alpha_{t-1} + R\eta_t, \qquad t = 1, \ldots, T, \tag{3.2}$$

where $\eta_t \sim WN(0, \sigma^2 Q)$. The appearance of the constant, σ^2, in the expressions for $Var(\xi_t)$ and $Var(\eta_t)$ is arbitrary. However, when $N = 1$, this is a natural representation for many models. While the introduction of σ^2 implies no loss in generality, it can lead to some simplification, since it appears neither in the updating nor the prediction equations. Thus although the filter is conditional on the parameters in T, R Q and h_t, σ^2 need not be known.

The Kalman filter equations are derived initially for the system (3.1) and (3.2), and extended to the general state space model in the last sub-section. The symbol a_t will be used to denote the MMSLE of α_t based on all the information up to, and including, the current observation, y_t. Thus a_t is the MMSLE of α_t at time t, whereas $a_{t/t-1}$ denotes the MMSLE of α_t at time $t - 1$.

Prediction

At time $t - 1$ all the available information is incorporated in a_{t-1}, the MMSLE of α_{t-1}. This has a covariance matrix which can be written as $\sigma^2 P_{t-1}$ where P_{t-1} is known.

The form of the transition equation, (3.2), suggests that the MMSLE of α_t at time $t - 1$ is given by

$$a_{t/t-1} = Ta_{t-1}. \tag{3.3}$$

Subtracting α_t from both sides of (3.3) yields

$$a_{t/t-1} - \alpha_t = T(a_{t-1} - \alpha_{t-1}) - R\eta_t. \tag{3.4}$$

The right-hand side of (3.4) has zero expectation, and so $a_{t/t-1}$ is u-unbiased. The covariance matrix of the estimation error may be obtained directly as

$$E[(a_{t/t-1} - \alpha_t)(a_{t/t-1} - \alpha_t)'] = \sigma^2 TP_{t-1}T' + \sigma^2 RQR'.$$

Thus

$$a_{t/t-1} - \alpha_t \sim \text{WS}(0, \sigma^2 P_{t/t-1}), \tag{3.5}$$

where

$$P_{t/t-1} = TP_{t-1}T' + RQR'. \tag{3.6}$$

Given that $a_{t/t-1}$ is the MMSLE of α_{t-1}, the MMSLE of y_t at time $t-1$ is clearly

$$\tilde{y}_{t/t-1} = z_t' a_{t/t-1}. \tag{3.7}$$

The associated prediction error is

$$v_t = y_t - \tilde{y}_{t/t-1} = z_t'(\alpha_t - a_{t/t-1}) + \xi_t. \tag{3.8}$$

Since $(\alpha_t - a_{t/t-1})$ and ξ_t both have zero expectation, it follows that $E(v_t) = 0$. Hence

$$\begin{aligned}
\text{Var}(v_t) = E(v_t^2) &= E[z_t'(\alpha_t - a_{t/t-1})(\alpha_t - a_{t/t-1})' z_t] \\
&\quad + E(\xi_t^2) + 2E[z_t'(\alpha_t - a_{t/t-1})\xi_t].
\end{aligned}$$

The expectation of the cross-product term is zero, so

$$\text{Var}(v_t) = \sigma^2 z_t' P_{t/t-1} z_t + \sigma^2 h_t = \sigma^2 f_t. \tag{3.9}$$

To summarise, (3.3) and (3.6) are the *prediction equations* for the state vector and its covariance matrix, while (3.8) is the error made in predicting y_t at time $t-1$.

Updating

The role of the updating equations is to incorporate the new information in y_t with the information already available in the optimal predictor, $a_{t/t-1}$. The problem is directly analogous to the one considered in the previous section, where the prior information in (2.5) was to be combined with the sample information of (2.1). The prior information is now contained in (3.5), while the sample consists of a single observation derived from the measurement equation, (3.1). Thus the augmented model is

$$\begin{pmatrix} a_{t/t-1} \\ y_t \end{pmatrix} = \begin{pmatrix} I \\ z_t' \end{pmatrix} \alpha_t + \begin{pmatrix} a_{t/t-1} - \alpha_t \\ \xi_t \end{pmatrix}. \tag{3.10}$$

The disturbance term has zero expectation and covariance matrix

$$\begin{aligned}
&E\left[\begin{pmatrix} a_{t/t-1} - \alpha_t \\ \xi_t \end{pmatrix} (a_{t/t-1}' - \alpha_t', \xi_t) \right] \\
&\qquad = \sigma^2 \begin{bmatrix} P_{t/t-1} & 0 \\ 0 & h_t \end{bmatrix}.
\end{aligned} \tag{3.11}$$

It follows from (2.10) and (2.11) that the MMSLE of α_t is given by

$$a_t = P_t(P_{t/t-1}^{-1} a_{t/t-1} + z_t y_t/h_t), \tag{3.12}$$

where

$$P_t = (P_{t/t-1}^{-1} + z_t z_t'/h_t)^{-1}. \tag{3.13}$$

Thus

$$a_t - \alpha_t \sim WS(0, \sigma^2 P_t). \tag{3.14}$$

The updating formula in (3.12) may be put in a different form using the matrix inversion lemma given in the Appendix. In terms of (A.1) $P_t = D$, $P_{t/t-1}^{-1} = A$, $z_t = B$ and $h_t = C$. Therefore, on equating (3.13) with (A.2), it follows that

$$P_t = P_{t/t-1} - P_{t/t-1} z_t z_t' P_{t/t-1}/f_t, \tag{3.15}$$

where

$$f_t = z_t' P_{t/t-1} z_t + h_t; \tag{3.16}$$

cf. (3.9).

Expression (3.15) provides an updating formula which does not require any matrix inversions. Furthermore, substituting (3.15) into (3.12) gives

$$\begin{aligned}
a_t &= (P_{t/t-1} - P_{t/t-1} z_t z_t' P_{t/t-1}/f_t)(P_{t/t-1}^{-1} a_{t/t-1} + z_t y_t/h_t) \\
&= a_{t/t-1} + P_{t/t-1} z_t(y_t/h_t - z_t' a_{t/t-1}/f_t - z_t' P_{t/t-1} z_t y_t/h_t f_t) \\
&= a_{t/t-1} + f_t^{-1} P_{t/t-1} z_t(y_t f_t/h_t - z_t' a_{t/t-1} - z_t' P_{t/t-1} z_t y_t/h_t).
\end{aligned}$$

Using the definition of f_t in (3.16) leads, on cancelling and re-arranging, to the state updating equation

$$a_t = a_{t/t-1} + P_{t/t-1} z_t(y_t - z_t' a_{t/t-1})/f_t. \tag{3.17}$$

Note the role played by the prediction error, $y_t - z_t' a_{t/t-1}$. This contains all the new information in y_t, and it is used to update $a_{t/t-1}$ via the *Kalman gain*. The Kalman gain is the $m \times 1$ vector, $P_{t/t-1} z_t/f_t$. Like P_t and $P_{t/t-1}$ it is independent of the y_t's, and so may be calculated in advance.

Expressions (3.15) to (3.17) are the basic *updating equations*. The above proof assumes that $h_t > 0$, since if this were not the case, (3.11) would be singular. However, by using the results in Theil (1971, pp. 282–7) it can be shown that the updating equations are still valid if $h_t = 0$. This is of some importance, because in some state space representations the measurement equation contains no noise. The equation obtained for an ARMA process, (1.8), provides an example.

A second point to note about the Kalman filter as a whole is that it is still valid if some of the elements in the state vector are fixed. This follows directly from the discussion of the previous section.

General Form of the Kalman Filter

The prediction and updating equations for the general state space model, (1.1) and (1.2), may be derived in exactly the same way as for the special case considered above. Given a_{t-1}, the MMSLE of α_{t-1} at time $t-1$, with

$$a_{t-1} - \alpha_{t-1} \sim \text{WS}(0, P_{t-1}), \tag{3.18}$$

the prediction equations are

$$a_{t/t-1} = T_t a_{t-1} \tag{3.19a}$$

and

$$P_{t/t-1} = T_t P_{t-1} T_t' + R_t Q_t R_t', \qquad t = 1, \ldots, T. \tag{3.19b}$$

The updating equations are given by

$$a_t = a_{t/t-1} + P_{t/t-1} Z_t' F_t^{-1} (y_t - Z_t a_{t/t-1}), \tag{3.20a}$$

and

$$P_t = P_{t/t-1} - P_{t/t-1} Z_t' F_t^{-1} Z_t P_{t/t-1}, \tag{3.20b}$$

where

$$F_t = Z_t P_{t/t-1} Z_t' + S_t H_t S_t', \qquad t = 1, \ldots, T. \tag{3.20c}$$

The prediction error

$$v_t = y_t - Z_t a_{t/t-1}, \qquad t = 1, \ldots, T, \tag{3.21}$$

is now an $N \times 1$ vector. It has zero mean and covariance matrix, F_t, i.e. $E(v_t) = 0$ and $E(v_t v_t') = F_t$. As in the single equation case, the prediction error plays a key role in updating the state vector. The 'correction' made to $a_{t/t-1}$ in (3.20a) is equal to the Kalman gain, $P_{t/t-1} Z_t' F_t^{-1}$, multiplied by v_t.

If prior information is available, i.e.

$$\alpha_0 \sim \text{WS}(a_0, P_0), \tag{3.22}$$

where a_0 and P_0 are known, the Kalman filter will yield the MMSLE of α_T based on all T observations. The question of obtaining a suitable specification for (3.22) will be pursued in the next section.

4. Maximum Likelihood Estimation

Consider a state space model of the form (3.1) and (3.2), and suppose that every element in α_t is stochastic. Suppose also that the disturbances, ξ_t and η_t, are normal and that

$$\alpha_0 \sim N(a_0, \sigma^2 P_0), \tag{4.1}$$

where a_0 and P_0 are known. Given these assumptions, the $T \times 1$ vector of observations, y, will have a multivariate normal distribution with mean μ and covariance matrix, $\sigma^2 V$. The starting values, a_0 and P_0, together with the parameters in the transition and measurement equations, imply a particular specification of μ and V. In general, however, there will be no reason to evaluate either μ or V. The systematic application of the Kalman filter will yield the MMSE of y_t, given the previous observations, for $t = 1, \ldots, T$, together with the corresponding prediction error, ν_t. The likelihood function, (1.2.15) may therefore be obtained directly from the prediction error decomposition, (1.2.21).

The above set-up is usually approached from the opposite direction. A time series is assumed to be generated by a particular model, and this in turn implies a likelihood function, (1.2.15). Casting the model in state space form then provides a means of evaluating the likelihood function via the prediction error decomposition. The only problems which need to be solved are the determination of a suitable state space representation and the construction of starting values.

Two categories of state space model may be distinguished, depending on whether or not the state vector is stationary. If it is, the distribution of the initial state vector is automatically available.

The discussion below will be restricted to models with a state space representation given by (3.1) and (3.2). The main principles extend straightforwardly, with the Kalman filter yielding the multivariate decomposition, (1.2.27), in the general case (1.1), (1.2).

Stationary Models

If α_t is to be stationary, the transition matrix, T, in (3.2) must satisfy the conditions imposed on Φ in the multivariate AR(1) model, (1.1.15). Since (3.2) contains no constant term, this implies that $E(\alpha_t) = 0$, although this condition need not be restrictive. The matrix $\sigma^2 P_0$ is simply the unconditional covariance matrix of α_t, and so P_0 may be obtained by making use of the results in Section 2.7.

In terms of the notation of (2.7.12), $\Omega = \sigma^2 RQR'$. Thus

$$\text{vec}(P_0) = [I - T \otimes T]^{-1}\,\text{vec}(RQR'). \tag{4.2}$$

The reason for the introduction of the constant term, σ^2, will become apparent in the examples below. The advantage from the point of view of ML estimation is that it can be concentrated out of the likelihood function.

Example 1   The transition equation for an MA(1) model is given by (1.7). The initial state vector is $a_0 = a_{1/0} = 0$. Since $\alpha_t = (y_t, \theta\epsilon_t)'$, the initial matrix $P_0 = P_{1/0}$ may be obtained directly, i.e.

$$P_{1/0} = P_0 = \sigma^{-2}E(\alpha_t\alpha_t') = \begin{bmatrix} 1 + \theta^2 & \theta \\ \theta & \theta^2 \end{bmatrix}. \tag{4.3}$$

The first prediction error is $\nu_1 = y_1$, while $f_1 = 1 + \theta^2$. Application of the updating formulae gives

$$a_1 = \begin{pmatrix} y_1 \\ \theta y_1/(1 + \theta^2) \end{pmatrix} \quad \text{and} \quad P_1 = \begin{pmatrix} 0 & 0 \\ 0 & \theta^4/(1 + \theta^2) \end{pmatrix}. \tag{4.4}$$

The prediction equations for α_2 are

$$a_{2/1} = \begin{pmatrix} y_1\theta/(1 + \theta^2) \\ 0 \end{pmatrix} \tag{4.5}$$

and

$$\begin{aligned} P_{2/1} &= \begin{pmatrix} \theta^4/(1 + \theta^2) & 0 \\ 0 & 0 \end{pmatrix} + \begin{pmatrix} 1 & \theta \\ \theta & \theta^2 \end{pmatrix} \\ &= \begin{pmatrix} (1 + \theta^2 + \theta^4)/(1 + \theta^2) & \theta \\ \theta & \theta^2 \end{pmatrix}, \end{aligned} \tag{4.6}$$

and so

$$\nu_2 = y_2 - \theta y_1/(1 + \theta^2) \quad \text{and} \quad f_2 = (1 + \theta^2 + \theta^4)/(1 + \theta^2). \tag{4.7}$$

On repeating this process further, it will be seen that the Kalman filter is effectively computing the prediction errors from the recursion

$$\nu_t = y_t - \theta\nu_{t-1}/f_{t-1}, \qquad t = 1, \ldots, T, \tag{4.8}$$

where $\nu_0 = 0$, and

$$f_t = 1 + \theta^{2t}/[1 + \theta^2 + \cdots + \theta^{2(t-1)}]. \tag{4.9}$$

Example 2   If an ARMA process is observed with measurement error, as in (1.9), the Kalman filter can be used to construct the likelihood function given the ARMA parameters and h. The initial conditions are computed in the same way as for an ARMA process observed without error.

   An alternative approach to this problem would be to model y_t as a more general ARMA process. For example, as noted in Section 1.5, an $AR(p)$ process observed with measurement error becomes $ARMA(p, p)$. The advantages of allowing for the error component explicitly are twofold. Firstly, there is a reduction in the number of parameters, and secondly estimates of the underlying variables, w_t, can be extracted from the observations, y_t. This is carried out by smoothing, a technique which is described in the next section.

Example 3   Suppose that y_t is the sum of two *unobserved components*, w_{1t} and w_{2t}, which follow independent ARMA processes. The measurement equation is then

$$y_t = w_{1t} + w_{2t}, \tag{4.10}$$

while the state vector, $(\alpha'_{1t}\alpha'_{2t})'$, is made up of state vectors appropriate to the individual ARMA processes. Using obvious notation the initial conditions are:

$$a_0 = \begin{pmatrix} 0 \\ 0 \end{pmatrix}, \quad P_0 = \begin{pmatrix} P_0^{(1)} & 0 \\ 0 & P_0^{(2)} \end{pmatrix}. \tag{4.11}$$

Non-Stationary Models

If α_t is non-stationary, starting values can generally be constructed from the observations themselves. When $N = 1$, a convenient approach is often to form starting values from the first m observations. The recursions can begin at $t = 0$, with $a_0 = 0$ and $P_0 = \kappa I$, where κ is a large number, but in forming the likelihood function, only prediction errors for $t > m$ should be employed. An example of the application of this technique will be found in Section 7.3 where the random-walk parameter regression model is discussed.

5. Prediction, Smoothing and the Steady State

Formulating one-step ahead predictions is an integral part of the Kalman filter solution to the estimation problem. However, once estimates at time $t = T$ have been obtained, it is frequently necessary to make predictions more than one period ahead. Thus the general multi-step prediction problem is to obtain optimal estimates of y_{T+l} for $l \geqslant 1$, given a_T and P_T. By way of contrast, smoothing concerns the calculation of the best estimates of α_t for $t = 1, \ldots, T$, given all the observations. This is accomplished by a set of recursive equations which begin with a_T and $P_{T/T}$ and work backwards. Multi-step prediction and smoothing are considered in the next two sub-sections. The final topic concerns the concept of a steady state solution. The discussion throughout is in terms of the general state space model, (1.1), (1.2), unless otherwise stated.

Multi-Step Prediction

The relevant formulae for multi-step prediction may be found by extending the one-step ahead prediction equations given in (3.19). The updating equations are simply by-passed, and so the MMSLE of the state vector l periods ahead is given by the recursion

$$a_{T+l/T} = T_{T+l} a_{T+l-1/T}, \qquad l = 1, 2, \ldots, \qquad (5.1)$$

where $a_{T/T} = a_T$. The associated MSE matrix is obtained from

$$P_{T+l/T} = T_{T+l} P_{T+l-1/T} T'_{T+l} + R_{T+l} Q_{T+l} R'_{T+l},$$

$$l = 1, 2, \ldots, \qquad (5.2)$$

with $P_{T/T} = P_T$. The MMSLE of y_{T+l} is

$$\tilde{y}_{T+l/T} = Z_{T+l} a_{T+l/T}, \qquad (5.3)$$

the variance of the prediction error being given by

$$\mathrm{Var}(v_{T+l}) = \mathrm{MSE}(\tilde{y}_{T+l/T}) = Z_{T+l} P_{T+l/T} Z'_{T+l} + S_{T+l} H_{T+l} S'_{T+l}. \qquad (5.4)$$

Optimal predictors of future values of y_t, together with their associated MSE's may therefore be obtained very easily. No new programming of any substance is required. The theoretical development needed to solve the prediction problem is minimal, the solution being contained essentially within the Kalman filter solution to the estimation problem.

Example 1   In the AR(1) model for a single time series, $T_t = \phi$ and $y_t = \alpha_t$. Expression (5.1) may therefore be solved to yield

$$\tilde{y}_{T+l/T} = \phi^l y_T, \qquad l = 1, 2, \ldots \qquad (5.5)$$

If $\sigma^2 = 1$, the MSE is obtained from the recursion

$$P_{T+l/T} = \phi^2 P_{T+l-1/T} + 1, \qquad l = 1, 2, \ldots \qquad (5.6)$$

The one-step ahead prediction error is simply $\epsilon_{T+1} = y_{T+1} - \phi y_T$, and so $P_{T+1/T} = 1$. Repeated application of (5.6), then gives

$$P_{T+2/T} = \phi^2 + 1,$$

$$P_{T+3/T} = \phi^2(\phi^2 + 1) + 1 = \phi^4 + \phi^2 + 1,$$

and so on. In general

$$P_{T+l/T} = \phi^{2(l-1)} + \phi^{2(l-2)} + \cdots + \phi^2 + 1, \qquad l = 1, 2, \ldots$$

$$(5.7)$$

As $l \to \infty$, the MSE tends to $\sigma^2/(1 - \phi^2)$, which is simply the variance of the process. Thus if l is large, the future value is too far ahead for the current observation to yield any useful information.

Multi-step prediction may also be appropriate within the sample period, if there are *missing observations*. A likelihood function may then be constructed using the prediction error decomposition. Thus suppose the observation for an AR(1) model at time t is missing. No prediction error enters the likelihood function at time t, but $v_{t+1} = y_{t+1} - \phi^2 y_{t-1}$, while $f_{t+1} = 1 + \phi^2$. Further details will be found in Harvey and Pereira (1980) and Jones (1980).

Smoothing

Each step in the Kalman filter yields the MMSLE of α_t, given all current and past observations. The only estimator which utilises all the sample observations is a_T, the estimator of the state in the final period. The smoothing equations therefore begin from a_T and P_T and work backwards.

If $a_{t/T}$ and $P_{t/T}$ denote the smoothed estimator and its covariance matrix at time t, the smoothing equations may be written

$$a_{t/T} = a_t + P_t^*(a_{t+1/T} - T_{t+1}a_t), \qquad (5.8a)$$

and

$$P_{t/T} = P_t + P_t^*(P_{t+1/T} - P_{t+1/t})P_t^{*'} \qquad (5.8b)$$

where

$$P_t^* = P_t T_{t+1}' P_{t+1/t}^{-1}, \qquad t = T - 1, \ldots, 1, \qquad (5.8c)$$

with $a_{T/T} = a_T$ and $P_{T/T} = P_T$. Smoothing can be carried out at any point from $t = 2$ to T, but in statistical applications it makes sense to focus primarily on smoothing at time $t = T$.

A set of what may be termed *direct residuals* may be obtained from the smoothed estimators. These residuals are contained in the $N \times 1$ vector

$$e_t = y_t - Z_t a_{t/T}, \qquad t = 1, \dots, T, \tag{5.9}$$

which has zero mean and covariance matrix $Z_t P_{t/T} Z_t' + S_t H_t S_t'$. The direct residuals should not be confused with the *prediction error residuals*, ν_t, defined by (3.21). If the parameters in the state space model are known, the latter are uncorrelated over time, i.e. $E[\nu_t \nu_s'] = 0$ for $t \neq s$. This is a property which the direct residuals do not, in general, possess.

Example 2 Consider the simple model

$$y_t = \alpha_t + \xi_t, \qquad \xi_t \sim WN(0, \sigma^2), \tag{5.10a}$$

$$\alpha_t = \alpha_{t-1} + \eta_t, \qquad \eta_t \sim WN(0, \sigma^2 q), \tag{5.10b}$$

where the state, α_t, and the observation, y_t, are scalars. The state, which is known to follow a random-walk process, cannot be observed directly as it is contaminated by the unobservable disturbance, ξ_t. We therefore have a very simple example of what an engineer would call a 'signal plus noise' model. It is assumed that q, the signal–noise ratio, is known.

The prediction equations in this case reduce to

$$a_{t/t-1} = a_{t-1} \quad \text{and} \quad P_{t/t-1} = P_{t-1} + q, \tag{5.11a \& 5.11b}$$

while the updating equations are

$$a_t = a_{t/t-1} + P_{t/t-1}(y_t - a_{t/t-1})/(P_{t/t-1} + 1) \tag{5.12a}$$

and

$$P_t = P_{t/t-1} - P_{t/t-1}^2/(P_{t/t-1} + 1). \tag{5.12b}$$

Suppose that $T = 4$ observations, y_1, y_2, y_3 and y_4, are available and that α_0 has a mean of a_0 and covariance matrix, $\sigma^2 P_0$. The values of the observations are given in table 4.1, while $a_0 = 4$, $P_0 = 12$ and $q = 4$.

The MMSLE of α_1, given y_1, is

$$a_1 = 4 + (12 + 4)(4.4 - 4.0)/(12 + 4 + 1) = 4.376,$$

while

$$P_1 = 16 - 16^2/17 = 0.941.$$

Since $z_t = 1$ in the measurement equation, the MMSLE of y_t is equal to $a_{t/t-1}$ for all t; hence $\bar{y}_1 = a_1 = 4.376$. Repeating the calculations for $t = 2$, 3 and 4, then gives the results shown in table 4.1.

Table 4.1   Smoothed Estimators and Residuals

t	1	2	3	4
y_t	4.4	4.0	3.5	4.6
a_t	4.376	4.063	3.597	4.428
P_t	0.941	0.832	0.829	0.828
v_t	0.400	−0.376	−0.563	1.197
$a_{t/T}$	4.306	4.007	3.739	4.428
$P_{t/T}$	0.785	0.710	0.711	0.828
e_t	0.094	0.007	−0.261	0.172

The final estimates obtained are $a_4 = 4.428$ and $P_4 = 0.828$. These values may now be used as starting values in the smoothing algorithm (5.8). Equation (5.8a) reduces to

$$a_{t/T} = a_t + P_t P_{t+1/t}^{-1}(a_{t+1/T} - a_t), \qquad t = T - 1, \ldots, 1,$$

thus

$$a_{3/4} = 3.597 + 0.829(4.428 - 3.597)/4.829 = 3.739,$$

while

$$P_{3/4} = 0.829 + (0.828 - 4.829)(0.829/4.829)^2 = 0.711.$$

Both the direct and prediction error residuals have been calculated, the formulae (5.9) and (3.21) reducing to $e_t = y_t - a_{t/T}$ and $v_t = y_t - a_{t-1}$ respectively.

Steady-State Solutions

If Z_t, S_t, T_t, R_t, H_t and Q_t are time invariant, the Kalman filter will tend towards a steady state, in which P_t and $P_{t/t-1}$ are also time-invariant. A discussion of the general conditions under which this takes place will be found in Jazwinski (1970) and Anderson and Moore (1979). The two examples here are designed simply to illustrate the concept.

Example 3   An examination of P_t in table 4.1, shows that the system in (5.10) is rapidly approaching a steady state. Combining the prediction and updating equations (5.11b) and (5.12b) yields

$$P_t = P_{t-1} + q - (P_{t-1} + q)^2/(P_{t-1} + q + 1). \tag{5.13}$$

Setting $P_t = P_{t-1} = \bar{P}$ and solving the resulting quadratic gives $\bar{P} = 0.828$; cf. the value of P_4. Once a steady state is reached, the updating and prediction equations for P_t become redundant.

Example 4   For the MA(1) model considered in Example 1 of
Section 4, the matrix $P_{t/t-1}$ is given by

$$P_{t/t-1} = \begin{bmatrix} f_t & \theta \\ \theta & \theta^2 \end{bmatrix}, \qquad t = 1, 2, \ldots \tag{5.14}$$

cf. (4.3) and (4.6). If $|\theta| \leq 1$, it is clear from (4.9) that

$$\lim_{t \to \infty} f_t = 1. \tag{5.15}$$

Hence the recursions tend towards a steady state. The implications
of this will be explored in the next chapter.

Appendix   Matrix Inversion Lemma

Let D be an $n \times n$ matrix defined by

$$D = [A + BCB']^{-1}, \tag{A.1}$$

where A and C are nonsingular matrices of order n and m
respectively, and B is $n \times m$. Then

$$D = A^{-1} - A^{-1}B[C^{-1} + B'A^{-1}B]^{-1}B'A^{-1}. \tag{A.2}$$

The result may be verified directly by showing that $DD^{-1} = I$. Let
$E = C^{-1} + B'A^{-1}B$. Then

$$\begin{aligned} D^{-1}D &= I - BE^{-1}B'A^{-1} + BCB'A^{-1} - BCB'A^{-1}BE^{-1}B'A^{-1} \\ &= I + [-BE^{-1} + BC - BCB'A^{-1}BE^{-1}]B'A^{-1} \\ &= I + BC[-C^{-1}E^{-1} + I - B'A^{-1}BE^{-1}]B'A^{-1} \\ &= I + BC[I - EE^{-1}]B'A^{-1} = I. \end{aligned}$$

An important special case arises when B is an $n \times 1$ vector, b, and
$C = 1$. Then

$$(A + bb')^{-1} = A^{-1} - \frac{A^{-1}bb'A^{-1}}{1 + b'A^{-1}b}. \tag{A.3}$$

A number of other useful results on matrix inversion will be found in
Jazwinski (1970, pp. 261–2).

Notes

The seminal paper on state space is Kalman (1960), but the discussion in
Sections 2 and 3 follows the approach of Duncan and Horn (1972). On the

prediction error decomposition and ML estimation, see *inter alia*, Schweppe (1965) and Harvey and Phillips (1976a; 1979).

An interesting approach to Bayesian forecasting based on the Kalman filter is set out in Harrison and Stevens (1976).

Exercises

1. Consider the model

$$y_t = 4 + \alpha_t + \xi_t, \qquad \xi_t \sim WN(0, \sigma^2)$$
$$\alpha_t = 0.5\alpha_{t-1} + \eta_t, \qquad \eta_t \sim WN(0, 4\sigma^2).$$

If $T = 4$ and the observations are as in table 4.1, complete the remaining entries and evaluate the likelihood function.

Find an expression for the steady state covariance matrix, $P_{t/t-1}$.

2. Suppose the transition equation (3.2) is modified to

$$\alpha_t = T_t \alpha_{t-1} + B_t x_t + \xi_t, \qquad t = 1, \ldots, T,$$

where x_t is a vector of known variables and B_t is a matrix of known coefficients. How are the Kalman filtering equations affected?

3. Consider the state space model with measurement equation (5.10a) and transition equation

$$\alpha_t = (-1)^{2t-1}\alpha_{t-1}, \qquad t = 1, 2, 3, \ldots$$

with $\alpha_0 \neq 0$. Show that the steady state matrix $P_t = \bar{P}$ is zero. Comment on this result.

5
Estimation of Autoregressive-Moving Average Models

1. Introduction

This chapter deals with the estimation of ARMA(p, q) models of the form (2.1.25). The models are assumed to be correctly specified in the sense that p and q are known. Problems associated with determining suitable values of p and q and assessing the consequences of misspecification will be considered in the next chapter.

It will be assumed throughout that the observations are normally distributed. The principle of maximum likelihood estimation then leads to procedures which are essentially based on minimising a residual sum of squares function. This is very convenient from the point of view of estimation, although, if the normality assumption is in doubt, it leaves the resulting estimators open to criticisms concerning lack of robustness; cf. the discussion in EATS (Section 3.7).

In discussing the large sample properties of estimators in Section 4 it will be assumed that the true model is both stationary and invertible. These assumptions will be retained in Section 5, when formal test procedures are derived. However, it is important to note that models in which some of the roots of the MA polynomial lie on the unit circle are perfectly valid and may arise in practice when overdifferencing has taken place. As a simple example suppose that $y_t = \epsilon_t$ where $\epsilon_t \sim NID(0, \sigma^2)$. Then

$$\Delta y_t = \epsilon_t + \theta \epsilon_{t-1}, \tag{1.1}$$

where $\theta = -1$, and so Δy_t is generated by an MA(1) process which is, strictly speaking, non-invertible. Thus although it is sensible to demand that estimates of the AR parameters should satisfy the stationarity conditions, the same is not necessarily true with regard to invertibility. This matter is discussed further in Section 3, while the implications for model selection are taken up in the next chapter.

The exact likelihood function is constructed from the prediction error decomposition of Section 1.2. This represents something of a departure from the usual practice. However, the prediction error decomposition is a natural approach from a theoretical point of view. Apart from clarifying the structure of the exact likelihood function, it makes the nature of the approximation embodied in the conditional likelihood function fully apparent. The conditional likelihood function is obtained by fixing certain initial disturbances and/or observations, and maximising this function is equivalent to minimising a corresponding sum of squares function. The resulting estimators, which are known as conditional sum of squares estimators, are generally taken to serve as adequate approximations to the exact ML estimators.

An alternative approach to maximum likelihood is to base estimates on the correlogram. For a pure AR model, the estimates derived from the correlogram are fully efficient. Although this is no longer the case for MA and mixed models, estimates obtained from the correlogram can provide useful starting values for an iterative optimisation procedure.

A final point concerns the assumption of a zero mean for the models considered. This is not particularly restrictive, and the introduction of a non-zero mean has no important implications for the construction of estimators of ϕ and θ, or for their large sample properties. Furthermore, the sample mean, $\bar{y}$, is an asymptotically efficient estimator of μ. This is demonstrated formally for an AR(1) model in the next section. The net result is that for any ARMA model, ϕ and θ may be estimated by putting the observations in deviation from the mean form.

2. Autoregressive Models

The likelihood function for an AR(1) process was derived in (2.2.26). The same technique may be adopted for an AR(p) process, the first step being to carry out a prediction error decomposition,

$$\log L(y) = \sum_{t=p+1}^{T} \log(y_t/y_{t-1}, \ldots, y_1) + \log L(y_p). \tag{2.1}$$

This expression is similar to (1.2.19). The first term on the right-hand side of (2.1) may be interpreted as the logarithm of the joint distribution of $\epsilon_{p+1}, \ldots, \epsilon_T$, while $L(y_p)$ is the joint distribution of the first p observations, $y_p = (y_1, \ldots, y_p)'$. If the covariance matrix of y_p is denoted by $\sigma^2 V_p$, the full log–likelihood function may be written as

$$\log L = -(1/2)T \log 2\pi - (1/2)T \log \sigma^2 - (1/2)\log |V_p|$$

$$- (1/2)\sigma^{-2} \left[y_p' V_p^{-1} y_p + \sum_{t=p+1}^{T} (y_t - \phi_1 y_{t-1} - \cdots \right.$$

$$\left. - \phi_p y_{t-p})^2 \right]. \quad (2.2)$$

Exact ML Estimation

The parameter σ^2 may be concentrated out of the likelihood function. However, the resulting function is still non-linear in ϕ, and ML estimation must be carried out by numerical optimisation. The matrix V_p^{-1} may be obtained directly, a number of methods being available; see Box and Jenkins (1976) or Galbraith and Galbraith (1974).

For the AR(1) model, setting $\partial \log L/\partial \phi = 0$ yields a cubic equation in ϕ. This has a unique real root in the range $[-1, 1]$. A method of calculating this root directly is given in Beach and MacKinnon (1978).

Regression

The estimation of AR models may be simplified considerably by regarding the first p observations, $y_1, \ldots, y_p$, as fixed. This provides a theoretical justification for dropping the last term in (2.1), with the result that maximising the likelihood function becomes equivalent to minimising the sum of squares function

$$S(\phi) = \sum_{t=p+1}^{T} (y_t - \phi_1 y_{t-1} - \cdots - \phi_p y_{t-p})^2. \quad (2.3)$$

The ML estimator of ϕ is therefore obtained by an OLS regression of y_t on its lagged values, $y_{t-1}, \ldots, y_{t-p}$.

In large samples it makes very little difference whether estimation is carried out by exact ML or by regression. In the AR(1) model, for example, it can be seen from (1.2.26) that the only distinguishing feature of the exact likelihood function is the inclusion of the terms involving y_1^2 and $\log(1 - \phi^2)$. These terms are swamped by the remainder of the likelihood function if T is at all large, and the asymptotic distribution of the estimators of ϕ and σ^2 is unaffected if they are omitted.

The introduction of a non-zero mean into an AR(1) model, as in (1.1.1), results in the sum of squares function becoming

$$S(\phi) = \sum_{t=2}^{T} [y_t - \mu - \phi(y_{t-1} - \mu)]^2.$$

Differentiating with respect to μ yields

$$\tilde{\mu} = \frac{\displaystyle\sum_{t=2}^{T} y_t - \tilde{\phi} \sum_{t=1}^{T-1} y_t}{(T-1)(1-\tilde{\phi})} \simeq \bar{y}.$$

Thus μ may be estimated independently of ϕ.

The Yule–Walker Equations

Estimates of the parameters in ϕ may also be obtained from the correlogram. For a pure AR(p) model, the autocorrelations are given by the pth order difference equation (2.2.37). Writing out this equation for $\tau = 1, \ldots, p$ gives

$$\begin{pmatrix} 1 & \rho(1) & \cdots & \rho(p-1) \\ \rho(1) & 1 & \cdots & \rho(p-2) \\ \vdots & \vdots & & \vdots \\ \rho(p-1) & \rho(p-2) & \cdots & 1 \end{pmatrix} \begin{pmatrix} \phi_1 \\ \phi_2 \\ \vdots \\ \phi_p \end{pmatrix} = \begin{pmatrix} \rho(1) \\ \rho(2) \\ \vdots \\ \rho(p) \end{pmatrix}.$$

$$(2.4)$$

These are known as the *Yule–Walker equations*. Replacing $\rho(\tau)$ by $r(\tau)$ yields a set of linear equations which may be solved directly to yield estimates of $\phi_1, \ldots, \phi_p$.

Example 1   For the AR(2) model

$$\hat{\phi}_1 = \frac{r(1)[1 - r(2)]}{1 - r^2(1)}, \quad \hat{\phi}_2 = \frac{r(2) - r^2(1)}{1 - r^2(1)}.$$

$$(2.5)$$

Apart from 'end effects', the estimators obtained in this way are the same as those obtained by regressing y_t on p lagged values of itself. The estimators obtained from the Yule–Walker equations therefore have the same asymptotic distribution as the ML estimators.

3. Moving Average and Mixed Processes

The likelihood function for any ARMA(p, q) process may be
constructed from the prediction error decomposition. If necessary,
the model may be cast in state space form, and the prediction errors
calculated by the Kalman filter. An example of this technique
applied to an MA(1) model was given in Section 4.4. However,
finding a way of computing the likelihood function is only the first
step in an algorithm, for the function must then be maximised with
respect to the elements in ϕ and θ. This may be very time-consuming,
particularly if $p + q$ is large.

  If certain assumptions are made about initial values of the
disturbances and/or the observations, a *conditional* likelihood
function is obtained. In the pure AR case this led to a linear ML
estimator of ϕ. For MA and mixed processes the ML estimator is
still non-linear, but the calculation of the likelihood function is
simplified considerably. Furthermore, analytic derivatives are
readily available and these can be important in improving the
efficiency of the optimisation procedure.

The Conditional Sum of Squares for the MA(1) Model

The MA(1) model,

$$y_t = \epsilon_t + \theta\epsilon_{t-1}, \qquad t = 1, \ldots, T, \tag{3.1}$$

provides the simplest illustration of the techniques used in estimating
models other than those which are purely autoregressive. Suppose,
for the moment, that θ is known. If, at time $t - 1$, ϵ_{t-1} were known,
the MMSE of y_t would be $\tilde{y}_{t/t-1} = \theta\epsilon_{t-1}$, and the associated
prediction error would be

$$\epsilon_t = y_t - \theta\epsilon_{t-1}. \tag{3.2}$$

Future prediction errors, each identical to the corresponding
disturbance term, could then be calculated from the recursion (3.2).
Unfortunately, there is a difficulty with this approach, namely that
in order to compute the full set of prediction errors, $\epsilon_1, \ldots, \epsilon_T$, the
initial disturbance, ϵ_0, must be known. One solution to the problem
is to assume that ϵ_0 is equal to zero. The full set of prediction errors
may then be computed, and if it is further assumed that ϵ_0 is fixed at
zero in all realisations, the log–likelihood function is of the form
(1.2.22). For most practical purposes, it makes little difference
whether the assumption $\epsilon_0 = 0$ is true or not. However, if ϵ_0 is
random, it is important to note that the approximation implicit in
the conditional likelihood is only valid if the process is invertible, i.e.

if $|\theta| < 1$. As long as this is the case, the difference between the exact and conditional likelihood functions becomes negligible as T increases.

Maximising the conditional likelihood function is equivalent to minimising the *conditional sum of squares* (CSS) function, $S(\theta) = \Sigma e_t^2$ with respect to θ. It is important to note that ϵ_t is no longer a disturbance term, but is a residual whose value depends on θ. The notation $\epsilon_t(\theta)$ stresses this point. However, it will usually be unnecessary to make this modification, since the interpretation of ϵ_t will be clear from the context.

For a given value of T, a set of T residuals may be generated from the recursion,

$$\epsilon_t = y_t - \theta\epsilon_{t-1}, \qquad t = 1, \ldots, T, \tag{3.3}$$

with $\epsilon_0 = 0$. These may then be used to construct $S(\theta)$ and the only remaining problem is to find a suitable method for minimising this function with respect to θ. Since only one parameter is involved, a grid search over the range $[-1, 1]$ could be carried out. For more general processes this approach may not be viable and the obvious algorithm to adopt is Gauss–Newton, or a suitable modification of it. For the MA(1) model, differentiating (3.3) yields

$$\frac{\partial\epsilon_t}{\partial\theta} = -\theta\,\frac{\partial\epsilon_{t-1}}{\partial\theta} - \epsilon_{t-1}, \qquad t = 1, \ldots, T. \tag{3.4}$$

Since $\epsilon_0 = 0$, it follows that $\partial\epsilon_0/\partial\theta = 0$. Thus the derivatives are produced by a recursion running parallel to (3.3), with the initialisation handled in a similar fashion. Given an estimate of θ, the algorithm proceeds by computing ϵ_t and $\partial\epsilon_t/\partial\theta$, and then updating the estimate from a regression of ϵ_t on $\partial\epsilon_t/\partial\theta$.

The Conditional Sum of Squares in the General Case

For higher order MA models, the conditional likelihood function is given by taking $\epsilon_{1-q}, \ldots, \epsilon_0$ to be equal to zero in all realisations. The residuals used to compute the CSS function are then obtained from the recursion

$$\epsilon_t = y_t - \theta_1\epsilon_{t-1} - \cdots - \theta_q\epsilon_{t-q}, \qquad t = 1, \ldots, T, \tag{3.5}$$

with $\epsilon_{1-q} = \epsilon_{2-q} = \cdots = \epsilon_0 = 0$.

Similar procedures may be adopted for mixed models, although in such cases there is the additional problem of handling the initial observations. Consider the ARMA(1, 1) model, (2.4.1). If y_1 is taken to be fixed, the prediction errors may be computed from the recursion

$$\epsilon_t = y_t - \phi y_{t-1} - \theta \epsilon_{t-1}, \qquad t = 2, \ldots, T \tag{3.6}$$

with $\epsilon_1 = 0$. An alternative approach would be to start the recursion at $t = 1$, with y_0 and ϵ_0 set equal to zero. However, although this yields T, rather than $T - 1$, residuals, it is not to be recommended, since arbitrarily setting $y_0 = 0$ introduces a distortion into the calculations. In general, the appropriate procedure for an ARMA(p, q) model is to compute $T - p$ prediction errors from a recursion of the form

$$\epsilon_t = y_t - \phi_1 y_{t-1} - \cdots - \phi_p y_{t-p} - \theta_1 \epsilon_{t-1} - \cdots - \theta_q \epsilon_{t-q},$$
$$t = p + 1, \ldots, T \tag{3.7}$$

with $\epsilon_p = \epsilon_{p-1} = \cdots = \epsilon_{p-q+1} = 0$.

The derivatives needed to implement the Gauss–Newton algorithm can again be computed by recursions. For the ARMA(1, 1) model,

$$\frac{\partial \epsilon_t}{\partial \phi} = -y_{t-1} - \theta \frac{\partial \epsilon_{t-1}}{\partial \phi}, \qquad t = 2, \ldots, T, \tag{3.8a}$$

$$\frac{\partial \epsilon_t}{\partial \theta} = -\theta \frac{\partial \epsilon_{t-1}}{\partial \theta} - \epsilon_{t-1}, \qquad t = 2, \ldots, T, \tag{3.8b}$$

with $\partial \epsilon_1 / \partial \phi = \partial \epsilon_1 / \partial \theta = 0$. In general, $p + q$ derivatives will be needed. However, these may be obtained on the basis of only two recursions, one for the AR and the other for the MA components; see Box and Jenkins (1976, p. 237).

Exact ML Estimation*

In calculating the conditional sum of squares for an MA(1) model, it is assumed that the initial disturbance, ϵ_0, is identically equal to zero. The traditional approach to evaluating the exact likelihood function in this case is to estimate ϵ_0 from the data. This requires at least two passes through the observations. An alternative procedure is to obtain the exact likelihood on the basis of the prediction error decomposition. By casting the model in state space form, the likelihood function can be computed by a single pass of the Kalman filter.

The state space representation of an MA(1) process was given in (4.1.7) and (4.1.8), while the evaluation of the likelihood function was discussed at some length in Section 4.4. An examination of the equations (4.4.4) to (4.4.9) is rather instructive, in that it provides a useful insight into the relationship between the conditional and exact likelihood functions. In both procedures the first prediction error is y_1, but in CSS the assumption that $\epsilon_0 = 0$ means that $\text{Var}(y_1) = \text{Var}(\epsilon_1) = \sigma^2$. On the other hand, if ϵ_0 is taken to be random, $\text{Var}(y_1) = \sigma^2(1 + \theta^2)$. This is reflected in f_1. Allowing for the distribution of ϵ_0 is also apparent in the second prediction error, (4.4.7), and its associated variance. In terms of the recursion in (4.4.8), the simplifying assumption in CSS means that f_{t-1} is effectively set equal to unity for all t.

It was observed in (4.5.14) that for an MA(1) model with $|\theta| \leqslant 1$, f_t tends towards unity as t increases. Thus for a sufficiently large value of t, the prediction error recursion, (4.4.8), may be approximated by the CSS recursion (3.3). This feature is exploited in the algorithm by Gardner, Harvey and Phillips (1980), where f_t is monitored. Once it becomes 'reasonably close' to unity, the programme switches to evaluating the prediction errors from a recursion of the form (3.3). A small number, δ, defines what is meant by 'reasonably close' in this context, the switch being made when $f_t < 1 + \delta$. Making δ smaller has the effect of diminishing the error in the approximation, but at the cost of increasing t^*, the value of t at which the switch takes place. A guide to the choice of δ is given in Gardner, Harvey and Phillips (1980) where table I shows the trade-off between t^* and the error of approximation. Setting $\delta = 0.01$ appears to be a reasonable compromise.

Another point to emerge from (4.4.9) is that (4.5.14) no longer holds when $|\theta| > 1$ since

$$\lim_{t \to \infty} f_t = \theta^2 > 1, \qquad |\theta| > 1. \tag{3.9}$$

Non-invertible solutions will therefore be detected by monitoring f_t. Note that when $|\theta| = 1$, in which case the process is, strictly speaking, non-invertible, f_t does converge to unity as $t \to \infty$. This is reasonable since $|\theta| = 1$ corresponds to a perfectly valid MA process; cf. (1.1).

The Kalman filter algorithm is applicable to any ARMA(p, q) model. However, computing the exact likelihood will always be more time-consuming than computing the conditional sum of squares. Furthermore, the fact that analytic derivatives are readily available in the CSS case means that numerical optimisation can, as a rule, be carried out more efficiently. Nevertheless, exact ML estimation may have some statistical advantages, particularly if the MA parameters lie close to, or on, the boundary of the invertibility region. The evidence for this assertion is presented in Section 6.

Initial Estimates

Starting values are needed for the Gauss—Newton iterations. For an MA(1) model it would be possible to begin by setting $\hat{\theta} = 0$, but a similar strategy could not be adopted for a mixed model. Setting $\hat{\phi} = \hat{\theta} = 0$ in the ARMA(1, 1) model would result in the two derivatives being identical. Hence the algorithm would immediately break down due to perfect multicollinearity in the regression of ϵ_t on $\partial \epsilon_t / \partial \phi$ and $\partial \epsilon_t / \partial \theta$.

A better way of starting the iterations is to begin from consistent estimates of the parameters. One possibility is to obtain estimates from the correlogram. Consider the MA(1) model. If $\rho(1)$ is replaced by $r(1)$ in (2.1.15), an estimator of θ is obtained by solving the quadratic equation,

$$\hat{\theta}^2 r(1) - \hat{\theta} + r(1) = 0. \tag{3.10}$$

This has two solutions, but since

$$r^{-1}(1) = \hat{\theta} + \hat{\theta}^{-1},$$

one root is clearly the reciprocal of the other. The problem of deciding which root to select is resolved by the invertibility condition. The estimator is therefore

$$\hat{\theta}_c = \{1 - [1 - 4r^2(1)]^{1/2}\}/2r(1). \tag{3.11}$$

The other solution, $\hat{\theta}_c^{-1}$, is ruled out, since it will have an absolute value greater than one.

The only case when (3.10) does not have two different solutions is when $r(1) = \pm 0.5$. The estimator is then $\hat{\theta}_c = \pm 1$. If $|r(1)| > 0.5$, there is no real solution to (3.10), since the theoretical first-order autocorrelation cannot exceed 0.5 for an MA(1) model. An $r(1)$ of this magnitude would probably suggest that an MA(1) model should not be entertained, although in small samples such a value could arise from an MA(1) process due to sampling fluctuations.

A similar technique may be employed for estimating the parameters in an ARMA(1, 1) model. The autocorrelation function is given by (2.4.13), which suggests estimating ϕ by

$$\hat{\phi}_c = r(2)/r(1). \tag{3.12}$$

If $\rho(1)$ and ϕ are replaced by $r(1)$ and $\hat{\phi}_c$ in (2.4.13a), an estimator is again given by the solution to a quadratic equation. This approach may be generalised to higher order models, although it does begin to get complicated; see Godolphin (1976). An alternative method for obtaining starting values has been developed by Wilson. This is described in Box and Jenkins (1976, pp. 201–5).

Although consistent estimators may be obtained from the correlogram, they will not be asymptotically efficient if the model contains an MA component. In the MA(1) case, for example, $\text{Eff}(\hat{\theta}_c) = 0.49$ when $\theta = \pm 0.5$, and as $|\theta| \to 1$, $\text{Eff}(\hat{\theta}_c) \to 0$. This implies that the higher order autocorrelations contain information which is not in $r(1)$.

Stationarity and Invertibility*

For the AR(1) process it is clear from (1.2.26) that the likelihood function is not defined for $|\phi| \geqslant 1$. As a result, the exact ML estimator will always lie within the unit circle. However, when estimators are computed by other methods, there is no longer any guarantee that they will satisfy the stationarity conditions. For example, Harvey and Phillips (1977) found that for $\phi = 0.9$, the regression estimator produced an estimate greater than unity in five out of two hundred replications.

In the case of higher order AR models and mixed models stationarity is not necessarily imposed on the estimates even when full ML is employed. Hence any estimates of the AR parameters should be checked against the stationarity condition, and if an iterative optimisation process is being employed, it may be wise to monitor the estimates produced at each step. An alternative approach is to employ a constrained optimisation procedure.

Violations of the invertibility conditions on the MA parameters have rather different implications. A non-invertible MA or ARMA process is still a stationary stochastic process with a properly defined likelihood function. However, allowing non-invertible estimates means that the likelihood equations have multiple solutions, and an identification problem arises. This is well illustrated by the MA(1) process. For every value of θ outside the unit circle, there is a value of θ within the unit circle with the same autocorrelation function. This follows because (2.1.15) takes the same value when θ is replaced by its reciprocal $1/\theta$. The implication is that the (exact) likelihood function will have two maxima, one at $\tilde{\theta}$ and one at $1/\tilde{\theta}$. The only exception to this rule is when the maximum is actually on the unit circle, i.e. $\tilde{\theta} = \pm 1$. Insisting that estimates of θ satisfy the invertibility conditions removes the ambiguity arising from multiple solutions. However, the fundamental reason for eliminating non-invertible estimates is that they lead to inefficient predictions; see Section 6.4.

There are basically two approaches to the problem of handling invertibility in full ML estimation. The first is to employ an unconstrained optimisation procedure. In the event of a maximum being located outside the invertibility region, the corresponding maximum within the invertibility region must then be determined. However, if such an approach is followed, it is important to compute the likelihood function by a method which does not suffer from rounding errors when non-invertible estimates arise during the iterations; see Osborn (1976). The Kalman filter algorithm and the related method proposed by Ansley (1979) are believed to be satisfactory in this respect. The second approach is to impose invertibility on the estimates from the beginning by applying a suitable constrained non-linear search routine, or by carrying out a transformation of the parameters; cf. EATS (Section 4.2). However, some care should be taken in the choice of transformation, since there is a tendency for exact ML estimates to be located right on the boundary of the invertibility region; see Section 6. For procedures other than full ML, the arguments for imposing the invertibility conditions on the estimates are much stronger. In particular, the CSS function has no meaning for MA parameters outside the invertibility region.

4. Asymptotic Properties

The asymptotic properties of ML estimators in ARMA models are usually established by examining the corresponding CSS estimators. Provided the model is both stationary and invertible, the difference between the exact and conditional likelihood functions is negligible in large samples, and so the asymptotic properties of the two sets of estimators are the same. For a pure AR model, minimising the CSS leads directly to the regression estimator. Hence the Mann–Wald theorem is applicable; cf. EATS (Section 1.4). For MA and mixed models no formal justification will be given for the assertion that the ML estimators are asymptotically normal with a covariance matrix equal to the inverse of the information matrix, and reference should be made to Fuller (1976), Hannan (1970) or Crowder (1976).

Autoregressive Processes

For an AR(1) model the information matrix may be evaluated directly by taking the expectation of the second derivatives of $\log L$. This yields

$$\text{Avar}(\tilde{\phi}, \tilde{\sigma}^2) = T^{-1} IA^{-1}(\phi, \sigma^2) = \begin{bmatrix} (1 - \phi^2)/T & 0 \\ 0 & 2\sigma^4/T \end{bmatrix}. \quad (4.1)$$

The result generalises to higher order models. In all cases the estimator of σ^2 is distributed independently of $\tilde{\phi}$ in large samples with variance $2\sigma^4/T$. The asymptotic variance of $\tilde{\phi}$ is

$$\text{Avar}(\tilde{\phi}) = T^{-1}V_p^{-1}, \tag{4.2}$$

where $\sigma^2 V_p$ is the covariance matrix of $(y_1, \ldots, y_p)'$.

MA and Mixed Processes

As in pure AR models, the ML estimator of σ^2 is distributed independently of the other estimators in large samples with a variance $2\sigma^4/T$. An expression for the asymptotic covariance matrix of the ARMA parameters, $\psi = (\phi', \theta')'$, may be obtained by evaluating the inverse of

$$IA(\psi) = \sigma^{-2} \text{ plim } T^{-1} \sum z_t z_t', \tag{4.3}$$

where $z_t = -\partial \epsilon_t / \partial \psi$, and dividing by T.

For the MA(1) model, $\partial \epsilon_t / \partial \theta$ obeys the recursion (3.4) and so z_t follows an AR(1) process,

$$z_t = (-\theta)z_{t-1} + \epsilon_{t-1}, \qquad t = 1, \ldots, T. \tag{4.4}$$

Hence

$$\text{plim } T^{-1} \sum_{t=1}^{T} z_t^2 = \text{Var}(z_t) = \sigma^2/(1 - \theta^2), \tag{4.5}$$

and so $\tilde{\theta}$ is asymptotically normally distributed with mean θ and variance

$$\text{Avar}(\tilde{\theta}) = (1 - \theta^2)/T. \tag{4.6}$$

Mixed models may be handled in the same way. In the ARMA(1,1) case, it follows from (3.8b) that $z_{2t} = -\partial \epsilon_t / \partial \theta$ is generated by an AR(1) process similar in form to (4.4). On the other hand, setting $z_{1t} = -\partial \epsilon_t / \partial \phi$ yields

$$z_{1t} = -\theta z_{1, t-1} + y_{t-1}.$$

This may be re-written as

$$z_{1t} = \frac{y_{t-1}}{1 + \theta L} = \frac{1}{1 + \theta L} \frac{1 + \theta L}{1 - \phi L} \epsilon_{t-1} = \frac{\epsilon_{t-1}}{1 - \phi L}$$

and so z_{1t} also obeys an AR(1) process,

$$z_{1t} = \phi z_{1, t-1} + \epsilon_{t-1},$$

with the same disturbance term as z_{2t}.

Expression (4.3) may be evaluated by taking note of (4.5), and writing

$$E(z_{1t}z_{2t}) = E\left[\left(\frac{\epsilon_{t-1}}{1-\phi L}\right)\left(\frac{\epsilon_{t-1}}{1+\theta L}\right)\right]$$

$$= E\{[\epsilon_{t-1} + \phi\epsilon_{t-2} + \phi^2\epsilon_{t-3} + \cdots]$$

$$\times [\epsilon_{t-1} + (-\theta)\epsilon_{t-2} + (-\theta)^2\epsilon_{t-3} + \cdots]\}$$

$$= \sigma^2[1 + (-\phi\theta) + (-\phi\theta)^2 + \cdots] = \sigma^2/(1 + \phi\theta).$$

The asymptotic covariance matrix of the ML estimator of ϕ and θ is therefore

$$\text{Avar}(\tilde{\phi}, \tilde{\theta}) = \frac{1}{T}\frac{1+\phi\theta}{(\phi+\theta)^2}\begin{pmatrix} (1-\phi^2)(1+\phi\theta) & -(1-\phi^2)(1-\theta^2) \\ -(1-\phi^2)(1-\theta^2) & (1-\theta^2)(1+\phi\theta) \end{pmatrix}.$$

$$(4.7)$$

5. Hypothesis Tests and Confidence Intervals

Tests of hypotheses relating to the ARMA parameters may be constructed from the principles laid out in Section 1.3. The classical framework assumes that the null hypothesis is nested within a more general model. Thus the unrestricted model is an ARMA(p, q) process in which the parameters are $\psi = (\phi'\,\theta')'$, while in the restricted model the null hypothesis $H_0: \psi = \psi_0$ is imposed. A test of an AR(1) model against an MA(1) model cannot therefore be carried out, although a test of AR(1) against ARMA(1, 1) is quite legitimate. However, discriminating between non-nested processes does play an important part in model selection, and a method of handling the problem, based on the Akaike Information Criterion, is described in Section 6.3.

A second, less obvious restriction must also be imposed on the hypotheses which can be tested in an ARMA model. Suppose that autoregressive and moving average terms are added simultaneously to a model. Thus, for example, the unrestricted model might be an ARMA(2, 2) process which under the null hypothesis becomes ARMA(1, 1). If the null hypothesis is true, the ARMA(2, 2) model has $\phi_2 = \theta_2 = 0$ and so it is not identifiable. This implies that any null hypothesis which imposes common factor restrictions on a model cannot be tested by standard procedures.

Wald and Likelihood Ratio Tests

The Wald statistic was defined in (1.3.4). If the model is estimated by Gauss–Newton, the natural large sample estimator of $I^{-1}(\tilde{\psi})$ is

$$\text{avar}(\tilde{\psi}) = \tilde{\sigma}^2 (\textstyle\sum z_t z_t')^{-1} \tag{5.1}$$

where $z_t = -\partial\epsilon_t/\partial\psi$ is evaluated at $\psi = \tilde{\psi}$. The regression methodology of Gauss–Newton suggests a modification of the above procedure, in which a test is based on the classical F-statistic in the regression of $\epsilon_t(\tilde{\psi})$ on $z_t(\tilde{\psi})$. The estimator of σ^2 has a divisor of $T - p - q$, thereby corresponding to the estimator s^2 in classical linear regression, while dividing W by m, the number of restrictions under H_0, converts the χ^2 statistic into an F. The use of the F-distribution should not be taken to imply that the test is exact. The F-test is mainly used as a matter of convenience, although the analogy with classical regression does suggest that it might be more accurate in certain circumstances. A test on a single parameter in the model can be based on the classical t-statistic. Again the test will not be exact. The terms 'asymptotic t-ratio' and 'asymptotic F-ratio' are often employed in this context.

> *Example 1*   In an AR(1) model the CSS estimator of ϕ is obtained directly by regressing y_t on y_{t-1}. The asymptotic t-ratio is
>
> $$\tilde{\phi}/(s/\sqrt{\textstyle\sum y_{t-1}^2}), \tag{5.2}$$
>
> where s^2 corresponds to the unbiased estimator of σ^2 in classical linear regression. A test based on (5.2) is asymptotically equivalent to a test which uses the square root of $(1 - \tilde{\phi}^2)/T$ as an estimate of the standard error. This is because
>
> $$\text{plim}(s^{-2}\textstyle\sum y_{t-1}^2/T) = 1/(1 - \phi^2). \tag{5.3}$$

> *Example 2*   If an MA(1) model is estimated by Gauss–Newton, a test on θ can be carried out on convergence. This is based on the 't-statistic' associated with $\partial\epsilon_t/\partial\theta$.

The LR test statistic is

$$\text{LR} = T\log(\tilde{\sigma}_0^2/\tilde{\sigma}^2) = T\log(SSE_0/SSE), \tag{5.4}$$

where SSE_0 and SSE are the residual sums of squares, $\sum\epsilon_t^2$, for the restricted and unrestricted models respectively. The analogy with classical linear regression again suggests a modified test in which

$$\text{LR*} = \frac{(SSE_0 - SSE)/m}{SSE/(T-p-q)} \tag{5.5}$$

is taken to have an F distribution with $(m, T - p - q)$ degrees of freedom under H_0.

Lagrange Multiplier Tests

Suppose the restricted form of the model has been estimated. An LM test of H_0 can then be based on a single iteration of Gauss–Newton. This leads to the TR^2 statistic, (1.3.6), which can be tested as a χ^2_m variate.

> *Example 3*   Suppose that an AR(1) model has been fitted, and that we wish to test for the addition of an MA component. In other words, the unrestricted model is ARMA(1, 1), while $H_0: \psi = \psi_0$ has $\psi_0 = (\phi, 0)'$. The residual function of an ARMA(1, 1) process is
>
> $$\epsilon_t = y_t - \phi y_{t-1} - \theta \epsilon_{t-1} \tag{5.6}$$
>
> and the derivatives of ϵ_t with respect to ϕ and θ are given by the recursive expressions (3.8a) and (3.8b). The LM procedure is based on a regression of $\epsilon_t(\tilde{\psi}_0)$ on its derivatives evaluated at $\tilde{\psi}_0$. Since $\theta = 0$ under H_0, this reduces to a regression of ϵ_t on y_{t-1} and ϵ_{t-1}. The test is carried out by treating TR^2 as a χ^2_1 variate. The 'modified LM test' takes the form of a t-test on the coefficient of ϵ_{t-1}.

> *Example 4*   If the null hypothesis in the previous example had been $H_0: \phi = \theta = 0$, the test would have broken down. Since $\partial \epsilon_t / \partial \theta = -\epsilon_{t-1} = -y_{t-1}$ when $\phi = \theta = 0$ and since $\partial \epsilon_t / \partial \phi = -y_{t-1}$, the two explanatory variables in the regression of ϵ_t on its derivatives are perfectly collinear. The fact that the LM statistic cannot be computed in this case is a reflection of the general point on common factor restrictions made at the beginning of this section.

Confidence Intervals

The most straightforward way of constructing a confidence interval for a single parameter is based on standard regression methodology. An estimate of the asymptotic standard error of a parameter will appear as part of a Gauss–Newton optimisation algorithm, and an approximate confidence interval may then be based on the t-distribution.

An alternative approach is based on finding a contour for the sum of squares surface. For a single parameter this will not necessarily give the same answer as the method described above, although any discrepancy is likely to be small. The advantage of the sum of squares approach is that it can be used to construct a confidence region for several parameters. Suppose that m parameters are being considered and that $\chi_m^2(\alpha)$ is the χ^2 significance value for a test of size α. An approximate $1 - \alpha$ confidence region is then bounded by the contour on the sum of squares surface for which

$$SSE(\psi) = SSE(\bar{\psi})[1 + \chi_m^2(\alpha)/m]. \tag{5.7}$$

Further details will be found in Box and Jenkins (1976, pp. 228–31).

6. Small Sample Properties

Although some work has been done in the pure AR case, few analytic results are available for the distribution of estimators of ARMA models. It is therefore necessary to rely on Monte Carlo studies to provide evidence on small sample properties. In this section an attempt is made to bring together some of the results reported in the literature, the survey being based primarily on the work of Ansley and Newbold (1980), Davidson (1981), Dent and Min (1978), Harvey and Phillips (1977), Kang (1975), Nelson (1974; 1976), and Nelson and O'Shea (1979).

Autoregressive Processes

For the AR(1) process, the evidence suggests that unless T is very small, say less than 20, the difference between the full ML and the least squares estimator is likely to be negligible. In both cases there is a downward bias, which is approximately equal to $2\phi/T$; see Shenton and Johnson (1965).

While there is little to choose between the various asymptotically equivalent estimators in the AR(1) case, it appears that this is no longer true for higher order models. For $T = 100$ and $p = 2$ and 3, Dent and Min (1978) report sample MSE's for the ML estimates which, in some cases, are appreciably smaller than the MSE's obtained by other methods. As might be expected, the differences between the various procedures become more marked when the true parameters are close to the boundary of the stationarity region.

Confidence intervals and 't-statistics' can be constructed very easily when AR models are estimated by regression, cf. Example 5.1. Unless the true parameters are near to the stationarity boundary, these statistics appear to be quite reliable; see Nelson (1976) and Ansley and Newbold (1980).

Moving Average and Mixed Processes

The implications of invertibility for exact ML estimation were discussed at some length in Section 4 and it was argued that optimisation could be carried out without imposing constraints. In the MA(1) case, an ML estimate of θ outside the unit circle would simply be converted to an invertible estimate by taking its reciprocal. The CSS function, on the other hand, has no meaning for values of θ outside the unit circle. However, in small samples it may sometimes be the case that the CSS function has no minimum within the invertibility region, with the result that an unconstrained optimisation procedure may not converge or, if it does converge, will produce meaningless estimates. While it must be stressed that such behaviour is only likely with $|\theta|$ close to one and T small, it nevertheless points to the importance of applying a constrained non-linear search routine; the 'Box Complex algorithm' is recommended by Dent and Min (1978).

Allowing the full likelihood function to be maximised without imposing restrictions reveals a rather interesting result. For small samples, and a true value of θ close to the unit circle, a relatively high number of global maxima are located at a value of θ *exactly* equal to plus or minus one. The effect of this is shown in figure 5.1, which is constructed from table 2 in Harvey and Phillips (1977). For $T = 25$ and $\theta = 0.9$, approximately half of the estimates computed in each of 200 replications form a typical 'bell-shaped' distribution centred on $\theta \simeq 0.85$. The remaining 103 estimated are exactly equal to unity; cf. figure 1 in Ansley and Newbold (1980). The notion that an estimator can take a particular value with a non-zero probability might appear rather implausible. However, Sargan and Bhargava (1980) have recently provided analytic evidence on this point. They set out a method for evaluating the probability that $\theta = \pm 1$ for given values of θ and T.

In the same set of replications, Harvey and Phillips also computed CSS estimates. In 32 cases the minimum was at $\theta = 1$, but only because estimates outside the unit circle were deemed to be inadmissible. Thus, unlike the ML case, these did not represent global solutions to the optimisation problem.

Given these results on the distribution of the ML estimator, care must be exercised in comparisons between estimators based on sample MSE's, since this does not capture the rather unusual bi-modality of θ in the distribution of the ML estimator. However,

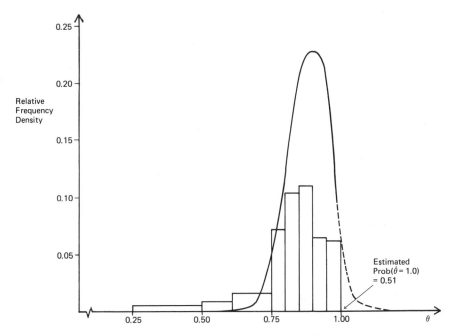

Figure 5.1   *Theoretical and Empirical Frequency Distributions for ML Estimator of θ in an MA(1) Process with θ = 0.9 (T = 25)*

most of the results quoted below are based on $T = 100$, and for a sample size of that order the effect is relatively unimportant unless θ is close to the unit circle. Nevertheless, the fact that a true value of θ equal to -1 can arise when overdifferencing takes place should be borne in mind when evaluating different estimators.

Nelson (1974), Davidson (1981) and Dent and Min (1978) compare a variety of estimation procedures over a range of values of θ. Nelson, unfortunately, does not include the ML estimator in his experiments, preferring to regard the 'unconditional sum of squares' as a reasonable approximation. The unconditional sum of squares estimator is obtained by maximising the exact likelihood function, but without the determinantal term, $\log |V|$. For a large value of T this estimator is similar to exact ML. However, if T is small and θ is close to the unit circle, it can be very different, and the evidence presented by Kang (1975) and Davidson (1981) suggests that there is little to recommend it. The main conclusions of the three studies are as follows. Firstly, for $T = 30$, the CSS estimator is clearly biased towards zero when $\theta = \pm 0.9$, but the extent of this bias is considerably reduced when $T = 100$. Secondly, the ML estimator is to be preferred to the CSS estimator on an MSE criterion when the true value of θ is close to the unit circle. However, when $T = 100$, the results of Dent and Min suggest that there is very little to choose between the two estimators for $|\theta|$ less than about 0.8. In fact, if anything, the CSS estimator has a smaller MSE. Supporting evidence is provided by Davidson.

Nelson also examines another aspect of inference, namely the estimation of standard errors. These may be computed from the cross-product matrix which emerges in the Gauss—Newton method, or by numerically evaluating the Hessian. For the CSS estimator, the estimated standard errors tend to be a good deal smaller than the corresponding MSE's given over all replications. The result is that the dispersion of 't-ratios' is too wide, and so a null hypothesis will tend to be rejected much too often. For $T = 30$, Nelson's results indicate that a nominal test size of 0.05 is likely to produce an actual test size of around 0.10 for most values of θ. Some improvement is observed as the sample size increases, but, as might be expected, values of θ close to the unit circle continue to give rise to problems in this respect. For example, with $T = 100$, a nominal test size of 0.05 implies an actual size of around 0.12 when $\theta = 0.9$. Nelson sums up these findings by observing that the t-ratios in MA models are likely to be 'treacherous' in small samples. Supporting evidence is provided in Nelson and O'Shea (1979) and Ansley and Newbold (1980).

The performance of estimators other than ML and CSS is also reported in some papers. Of principal interest is the correlogram estimator, $\hat{\theta}_c$, defined in (3.11). When θ is close to zero, $\hat{\theta}_c$ has an MSE close to that of the CSS estimator, but its relative efficiency declines fairly rapidly as $|\theta|$ tends towards one. Although its MSE diminishes as T increases, its relative performance with respect to CSS actually gets worse. In Nelson's study, the sample MSE of $\hat{\theta}_c$ is about twice that of the corresponding MSE for the CSS estimator when $\theta = 0.5$ and $T = 30$, but around three times the size for $T = 100$. The asymptotic efficiency of $\hat{\theta}_c$ is 0.49 for $\theta = 0.5$. The results given by Dent and Min for $T = 100$ are very similar. The overall conclusion is that although $\hat{\theta}_c$ might not be entertained as a final estimator, it is nevertheless a satisfactory way of generating starting values for iterative schemes. To quote Nelson (1974, p. 127) '. . . the GN method led to the global minimum (for CSS) reliably and quickly when $\hat{\theta}_c$ was used as an initial guess value.'

For the MA(2) process, the results in Dent and Min again indicate that there is little to choose between the ML and CSS estimators, except when the parameters are close to the boundary of the invertibility region. In such cases there appears to be a gain in using the ML estimator. The conclusion reached by Dent and Min (1978,

p. 38) that '. . . for pure moving average models the conditional least squares (CSS) estimator may be marginally superior to the maximum likelihood estimator' is not particularly well supported by their results for the MA(2) process. The MSE's quoted for the CSS estimator fall below the corresponding MSE's in less than half of the models studied, and the differences are so small that it is difficult to draw firm conclusions on the basis of only one hundred replications.

For the ARMA(1, 1) model, the results in Dent and Min show the ML and CSS estimators to be performing equally well; no clear pattern of superiority of one over the other emerges. The results in Ansley and Newbold (1980), on the other hand, indicate a clear gain from using full ML for both ϕ and θ. As in the MA(1) case, the 'pile up' effect of estimates of θ when the true value is close to the unit circle should be borne in mind in any comparison of MSE's. For $\theta = 0.9$ and $\phi = 0.9$, Harvey and Phillips (1977) found that, for $T = 25$, between 40% and 50% of the ML estimates of θ were exactly equal to unity. A similar proportion was observed when ϕ was set equal to 0.5.

Ansley and Newbold (1980) also studied the reliability of confidence intervals for mixed models. They found particularly severe problems when there was 'anything approaching parameter redundancy'. In such cases the computed confidence intervals can be far too narrow.

7. Multivariate Models

The general zero mean vector ARMA(p, q) process is

$$y_t = \Phi_1 y_{t-1} + \cdots + \Phi_p y_{t-p} + \epsilon_t + \Theta_1 \epsilon_{t-1} + \cdots + \Theta_q \epsilon_{t-q},$$
$$t = 1, \ldots, T \quad (7.1)$$

where y_t is an $N \times 1$ vector of observations and $\Phi_1, \ldots, \Phi_p$, $\Theta_1, \ldots, \Theta_q$ are the $N \times N$ matrices of ARMA parameters. The disturbance vector, ϵ_t, will be taken to be normally distributed and so $\epsilon_t \sim NID(0, \Omega)$, where Ω is an $N \times N$ p.d. matrix.

The methods available for estimating models of this form are basically generalisations of the procedures developed in the univariate case. Exact ML estimation may, in principle, be carried out as before. However, the extra parameters introduced to link the series together pose more complex computational problems. In some cases this means that exact ML estimation is ruled out as a practical proposition. Relatively more attention is therefore paid to the generalisation of the CSS procedure. The multivariate Gauss—Newton algorithm may be employed in the general case, and, as for univariate models, the efficiency of this approach is considerably enhanced by the fact that analytic derivatives are relatively easy to compute.

The discussion below will be confined to the vector AR(1) and MA(1) models. The first of these models is important insofar as it is the simplest multivariate process from the point of view of estimation. Once this model can be handled, the extension to higher order autoregressive processes is relatively straightforward. Similarly, the vector MA(1) process illustrates most of the difficulties involved in estimating the general multivariate ARMA(p, q) process.

Estimation of the Vector AR(1) Process

Because of its relative simplicity the first-order vector autoregressive process

$$y_t = \Phi y_{t-1} + \epsilon_t, \qquad t = 1, \ldots, T, \tag{7.2}$$

has received considerable attention, particularly in the econometric literature. However it still contains a large number of unknown parameters. The autoregressive matrix, Φ, contains N^2 parameters, while the symmetric covariance matrix, Ω, contributes a further $N(N+1)/2$ distinct terms.

The likelihood function for (7.2) may be derived exactly as in the scalar case. Given the vector y_{t-1}, the distribution of y_t is multivariate normal with a mean of Φy_{t-1} and a covariance matrix Ω, and so

$$\log L(y_T, y_{T-1}, \ldots, y_2/y_1) = -\frac{N(T-1)}{2} \log 2\pi$$

$$-\frac{1}{2}(T-1)\log|\Omega| - \frac{1}{2} \sum_{t=2}^{T} (y_t - \Phi y_{t-1})'\Omega^{-1}(y_t - \Phi y_{t-1}). \tag{7.3}$$

This is the log–likelihood function conditional on the first observation, y_1. The full log–likelihood function is obtained by taking account of the unconditional distribution of y_1. This is multivariate normal with a mean of zero and a covariance matrix, Σ, which is defined by

$$\Sigma = \Phi\Sigma\Phi' + \Omega; \tag{7.4}$$

cf. (2.7.11). Combining the distribution of y_1 with (7.3) yields the full log–likelihood function:

$$\log L(y_T, \ldots, y_1) = -\frac{NT}{2} \log 2\pi - \frac{1}{2} \log |\Sigma| - \frac{(T-1)}{2} \log |\Omega|$$

$$-\frac{1}{2} \sum_{t=2}^{T} (y_t - \Phi y_{t-1})' \Omega^{-1} (y_t - \Phi y_{t-1}) - \frac{1}{2} y_1' \Sigma^{-1} y_1. \quad (7.5)$$

The maximisation of (7.5) with respect to Ω and Φ may be carried out by a general numerical optimisation routine. However, it is considerably easier to treat y_1 as fixed, and work with the conditional log–likelihood function, (7.3). Given Ω, the ML estimator of Φ is then obtained by minimising

$$S = \sum_{t=2}^{T} (y_t - \Phi y_{t-1})' \Omega^{-1} (y_t - \Phi y_{t-1}). \quad (7.6)$$

However, this is precisely the criterion function which is minimised by the multivariate least squares estimator; see EATS (Section 2.10). In other words, the ML estimator of Φ does not depend on Ω, the ith row in $\tilde{\Phi}$ being given by a straightforward OLS regression of the ith element in y_t on the vector y_{t-1}, i.e.

$$\tilde{\Phi}' = \left[\sum_{t=2}^{T} y_{t-1} y_{t-1}' \right] \sum_{t=2}^{T} y_{t-1} y_t'. \quad (7.7)$$

The ML estimator of Ω may be calculated directly from the residuals, $e_t = y_t - \tilde{\Phi} y_{t-1}$, $t = 2, \ldots, T$. Thus

$$\tilde{\Omega} = T^{-1} \sum_{t=2}^{T} e_t e_t'. \quad (7.8)$$

Asymptotic Distribution of the ML Estimator and Associated Test Statistics

Let $\phi = \text{vec}(\Phi)$. The ML estimator of ϕ is then asymptotically normally distributed with mean ϕ and covariance matrix

$$\text{Avar}(\tilde{\phi}) = T^{-1} (\Omega \otimes \Sigma^{-1}). \quad (7.9)$$

The easiest way to estimate Σ is from the formula

$$\hat{\Sigma} = T^{-1} \sum_{t=2}^{T} y_t y_t'. \quad (7.10)$$

Tests of hypotheses concerning Φ may be carried out very easily using the Wald principle. The appropriate statistic for testing $H_0 : \Phi = 0$ against $H_1 : \Phi \neq 0$ is

$$W = \tilde{\phi}' [\mathrm{Avar}(\tilde{\phi})]^{-1} \tilde{\phi}. \tag{7.11}$$

When H_0 is true, W has a χ^2 distribution with N^2 degrees of freedom in large samples.

An alternative approach to testing hypotheses is to use the LR statistic. For a test of $\Phi = 0$,

$$\mathrm{LR} = T \log(|\hat{\Sigma}|/|\tilde{\Omega}|). \tag{7.12}$$

Wald and LR statistics may also be used to test whether Φ is diagonal. In fact it may be useful to first test Φ for diagonality and, if this hypothesis is not rejected, to then test whether the diagonal elements are zero. These hypotheses are nested, and so if W_1 is the Wald statistic for testing the diagonality of Φ, and W is the test statistic (7.11), W_1 and $W - W_1$ are asymptotically distributed as independent χ^2 variates with $N(N - 1)$ and N degrees of freedom respectively; see EATS (Section 5.8). Note that if Φ is taken to be diagonal, the ML estimates of Φ will no longer be independent of Ω, and so a sequence of Wald tests involves less computation than a corresponding sequence of LR tests.

The Vector MA(1) Process*

The vector MA(1) was defined in (2.7.4). If $\epsilon_t \sim NID(0, \Omega)$ and $\epsilon_0 = 0$, the log–likelihood function is of the form (1.2.28). Conditional on Φ, the ϵ_t's may be obtained from the recursion

$$\epsilon_t = y_t - \Theta \epsilon_{t-1}, \qquad t = 1, \ldots, T, \tag{7.13}$$

while Ω is estimated by a formula analogous to (7.8). The likelihood function may be maximised with respect to Θ by the multivariate Gauss–Newton algorithm; see EATS (Section 4.4) and Wilson (1973). Let $\hat{\theta}$ be an initial estimator of the $N^2 \times 1$ vector $\theta = \mathrm{vec}(\Theta)$. For a given value of Ω, this may be updated by a generalisation of the Gauss–Newton formula,

$$\theta^* = \hat{\theta} + \left[\sum_{t=1}^{T} Z_t \Omega^{-1} Z_t' \right]^{-1} \sum_{t=1}^{T} Z_t \Omega^{-1} \epsilon_t, \tag{7.14}$$

where Z_t is the $N^2 \times N$ matrix of derivatives defined by $Z_t = -\partial \epsilon_t'/\partial \theta$. This is repeated until convergence with the estimate of Ω updated after each iteration.

The derivatives may be calculated recursively just as in the univariate case. If the model is written in the form

$$y_t = \epsilon_t + (I \otimes \epsilon'_{t-1})\theta, \qquad (7.15)$$

the residual recursion, (7.13), becomes

$$\epsilon_t = y_t - [I \otimes \epsilon'_{t-1}]\theta \qquad (7.16)$$

and so

$$Z'_t = [I \otimes \epsilon'_{t-1}] + [I \otimes Z'_{t-1}]\theta, \qquad t = 1, \ldots, T \qquad (7.17)$$

with $Z_0 = 0$.

An estimator of the asymptotic covariance matrix of the ML estimator is given automatically by the multivariate Gauss–Newton algorithm. This is

$$\text{avar}(\tilde{\theta}) = \left[\sum_{t=1}^{T} Z_t \tilde{\Omega}^{-1} Z'_t \right]^{-1}, \qquad (7.18)$$

all the elements in Z_t being evaluated at $\theta = \tilde{\theta}$. Hence Wald tests can be carried out just as in the vector AR(1) model.

Seemingly Unrelated ARMA Processes*

Specifying diagonal coefficient matrices and employing a two-step estimation procedure leads to considerable simplification for MA and mixed processes. Consistent estimates of the ARMA parameters may be obtained by treating each of the N processes separately, while an estimate of Ω may be constructed from the residuals. One iteration of the multivariate Gauss–Newton scheme yields an asymptotically efficient estimator of the full set of ARMA parameters. This estimator has the form of a feasible GLS estimator in a system of seemingly unrelated regression equations.

The Monte Carlo experiments reported in Nelson (1976) examine the small sample gains which can be expected by treating individual ARMA processes as a system. For a vector AR(1) process with $N = 2$, i.e.

$$\begin{bmatrix} y_{1t} \\ y_{2t} \end{bmatrix} = \begin{bmatrix} \phi_1 & 0 \\ 0 & \phi_2 \end{bmatrix} \begin{bmatrix} y_{1,t-1} \\ y_{2,t-1} \end{bmatrix} + \begin{bmatrix} \epsilon_{1t} \\ \epsilon_{2t} \end{bmatrix}, \qquad (7.19)$$

the relative efficiency of the single series estimator of one of the parameters, say ϕ_1, may be obtained analytically for large samples. If $\hat{\phi}_1$ and $\tilde{\phi}_1$ denote the single series and joint ML estimators respectively, and ρ is the correlation between ϵ_{1t} and ϵ_{2t}, the relative efficiency of $\hat{\phi}_1$ is

$$\text{Eff}(\hat{\phi}) = \frac{\text{Avar}(\tilde{\phi})}{\text{Avar}(\hat{\phi})} = (1 - \rho^2)\left[1 - \frac{\rho^4(1 - \phi_1^2)(1 - \phi_2^2)}{(1 - \phi_1\phi_2)^2} \right]^{-1}. \qquad (7.20)$$

For samples of size 30, (7.20) gives a very good guide to the gains in efficiency actually obtained. As might be expected, the higher the correlation between the series, the greater the gain in efficiency from estimating the parameters jointly. Furthermore, on examining (7.20) further it can be seen that significant gains in efficiency are associated with large differences in the parameter values. As $|\phi_1 - \phi_2|$ goes to its maximum of 2, $\text{Eff}(\hat{\phi})$ tends to its minimum value of $1 - \rho^2$.

A similar analysis may be carried out for two MA(1) processes. Again Nelson's results show the gains in efficiency to be approximately as predicted by asymptotic theory, even though the small sample distribution of the MA parameters is much more dispersed than asymptotic theory would suggest.

Exercises

1. Derive an LM test for testing the hypothesis that a model is ARMA(1, 1) against the alternative that it is ARMA(1, 3). How would you carry out a modified LM test based on the F-distribution?

2. Show that the LM test of $\theta = 0$ in the ARMA(1, 1) model is equivalent to a test based on the first autocorrelation, $r(1)$, computed from the residuals from an AR(1) model, with $r(1)$ regarded as being asymptotically normal with mean zero and variance ϕ^2/T.

3. For the set of observations 3.6, 4.8, 5.0, 6.6 and 4.7, calculate the conditional sum of squares for an MA(2) process with $\theta_1 = -0.5$ and $\theta_2 = -0.2$.

4. If the sample autocorrelations in a series are $r(1) = 0.8$ and $r(2) = 0.5$, find asymptotically efficient estimates of the parameters in an AR(2) model. If the variance of the observations is 2.0, estimate the variance of the disturbance term, ϵ_t. Compute asymptotic standard errors for the estimates of ϕ_1 and ϕ_2.

5. Sketch the power spectrum of the AR(2) process in Question 4. Find the location of its maximum and comment.

6. Show that the Yule–Walker equations for an AR(2) model give approximately the same estimators of ϕ_1 and ϕ_2 as a regression of y_t on y_{t-1} and y_{t-2}. Find an expression for the asymptotic covariance matrix of these estimators in terms of ϕ_1 and ϕ_2.

7. Given the information $r(1) = 0.60$, $r(2) = 0.32$, $r(3) = 0.18$ and $\bar{y} = 7.2$, from a sample of size 100, calculate asymptotically efficient estimates of the parameters in an AR(1) model with a non-zero mean. If $c(0) = 4.0$, calculate asymptotic standard errors for these estimates by deriving the information matrix.

8. The following statistic has been proposed for testing the hypothesis that $\Phi = \Phi_0$ in the vector AR(1) model, (7.2):

$$\text{tr}(\tilde{\Phi} - \Phi_0)\left[\sum_{t=2}^{T} y_t y_t'\right](\tilde{\Phi} - \Phi_0)\tilde{\Omega}^{-1}.$$

This is tested as a χ^2 variate with N^2 degrees of freedom. Is this test related to any of those discussed in the text?

9. How would you construct LR, Wald and LM tests of the hypothesis that $\theta_0 = 0$ in the model $y_t = \phi y_{t-1} + \theta_0 + \epsilon_t + \theta_1 \epsilon_{t-1}$?

6
Model Building and Prediction

1. Introduction

The autoregressive-moving average models described in the previous chapters can be used as the basis for modelling a univariate time series. Although ARMA processes are primarily designed for stationary time series, it is straightforward to extend them to form a class which encompasses a wide range of non-stationary behaviour. This is done by differencing. An ARMA model is fitted, not to the raw observations, but to the first or second differences. The result is an autoregressive-*integrated*-moving average (ARIMA) process.

The development of an appropriate methodology for fitting ARIMA models is associated primarily with G. E. P. Box and G. M. Jenkins. Indeed, the whole approach is often referred to simply as the *Box—Jenkins method*. The core of this method is described in Section 3 and the last sub-section of Section 7. Basically, it consists of a model selection strategy which is divided into three parts. The first part consists of selecting a tentative model. This means deciding on an appropriate degree of differencing as well as fixing values for p and q. Box and Jenkins refer to this as model identification. The second stage in their approach is estimation, a topic which was covered fully in the previous chapter. Finally, the model is subject to diagnostic checking. The sequence of identification, estimation and diagnostic checking is seen as a cycle which is repeated until a satisfactory model has been obtained. In describing this procedure here a number of modifications and suggestions for improvement are made, and so the strategy which emerges should not be regarded as being identical to the one set out in Box and Jenkins (1976).

The rationale underlying the construction of ARIMA models is discussed in Section 5. The idea of differencing is related to the notion of a stochastic trend and the connection between local and global trends is explored in some detail. Similar concepts are involved in the treatment of seasonality. Section 6 describes methods of modelling a deterministic seasonal component and then goes on to show how a slowly changing seasonal pattern can be accommodated by differencing. A general class of models is defined in Section 7.

The way in which predictions are made for ARMA models is set out in Section 4. The important question of calculating prediction MSE's is also considered, and the techniques developed are shown to extend straightforwardly to seasonal ARIMA models in Sections 5 and 6. Multivariate ARIMA model building is considered in the final section.

2. Tests of Randomness

A successful model will aim to capture the systematic movements in the data. If this goal is achieved, then what remains, namely the residuals, should essentially be random. In other words, they should contain no predictable systematic component. This section introduces the basic considerations which arise in assessing the randomness of an observed time series, and in Section 3 these ideas are extended to testing the residuals from a fitted ARMA model.

Let $y_1, \ldots, y_T$ be a realisation from a white noise process with mean μ and variance σ^2. The theoretical autocorrelations for such a process are, by definition, identically equal to zero for all non-zero lags. A set of sample autocorrelations, on the other hand, will tend to show some departure from this pattern because of sampling error. Therefore, if an assessment of randomness is to be based on the sample autocorrelations, it is first necessary to investigate their sampling distribution. Fortunately, a very simple result is available for large samples, namely that for $\tau \neq 0$, the $r(\tau)$'s are normally and independently distributed with a mean of zero and a variance of $1/T$. This follows directly from the Mann–Wald theorem (EATS, pp. 48–9), since $r(\tau)$ is approximately equal to the coefficient of $y_{t-\tau}$ in the regression of y_t on $y_{t-\tau}$. Because the series is random, this coefficient has a limiting normal distribution when multiplied by $\sqrt{T}$, the mean being zero and the variance being

$$\sigma^2 (\text{plim } T^{-1} \sum y_{t-\tau}^2)^{-1} = 1.$$

Using the above result, the hypothesis that $\rho(\tau) = 0$ may be formally tested for any particular value of τ by treating $T^{1/2} r(\tau)$ as a standardised normal variate. At the 5% level of significance, the null hypothesis is rejected if $|T^{1/2} r(\tau)| > 1.96 \simeq 2.00$. However, such a test is only valid if the lag to be tested is specified in advance. This implies some *a priori* knowledge of the nature of the series. For example, with quarterly data a test of the significance of $r(4)$ would clearly be relevant. Except for the case of a seasonal effect, however, such *a priori* knowledge is likely to be the exception rather than the rule, and formal test procedures are generally restricted to the first-order autocorrelation, $r(1)$. For this reason some effort has been devoted to developing exact test procedures for hypotheses concerning $\rho(1)$. These are discussed in the next sub-section.

Notwithstanding the above remarks, it is nevertheless very useful to plot two lines on the correlogram, one at a height of $2/\sqrt{T}$ above the horizontal axis and the other at the same distance below. These may be used as a yardstick for assessing departures from randomness. If the underlying series is white noise, most of the sample auto-correlations will lie within these limits. Values outside are an indication of non-randomness, although for a white noise process, about one in twenty of the autocorrelations will be significant.

The von Neumann Ratio

The von Neumann Ratio is defined by

$$\text{VNR} = \frac{T}{T-1}\left[\sum_{t=2}^{T}(y_t - y_{t-1})^2 \Big/ \sum_{t=1}^{T}(y_t - \bar{y})^2\right], \tag{2.1}$$

where $\bar{y}$ is the sample mean. This statistic is closely related to $r(1)$. Writing $y_t - y_{t-1} = y_t - \bar{y} - (y_{t-1} - \bar{y})$, substituting in the numerator of (2.1), and expanding shows that

$$\text{VNR} \simeq 2[1 - r(1)]. \tag{2.2}$$

The approximation arises simply because of the treatment of the end point. Its effect is negligible in moderate or large samples.

If the observations are normally distributed white noise, i.e. if $y_t \sim NID(0, \sigma^2)$, the small sample distribution of (2.1) is known. An exact test may therefore be carried out using the significance values tabulated by Hart (1942). From (2.2) it can be seen that when $r(1) = 0$, the von Neumann Ratio is approximately equal to two, but as $r(1)$ tends towards one, VNR tends towards zero. A one-sided test against positive serial correlation is therefore carried out by looking up the appropriate significance point and rejecting the null hypothesis if VNR falls below it. Conversely, a test against negative serial correlation is based on the upper tail of the distribution. Anderson (1971, Chapter 6) shows that against a one-sided alternative, the von Neumann Ratio test is uniformly most powerful unbiased.

If T is large, the distribution of VNR may be approximated by a normal distribution with a mean of $2T/(T-1)$ and a variance $4/T$. This may be used as the basis for a test procedure, although what amounts to the same test may be carried out directly on $r(1)$.

The Portmanteau Test Statistic

The result on the asymptotic distribution of the sample auto-correlations suggests that a general test of randomness might be based on the first P autocorrelations, through the statistic

$$Q = T \sum_{\tau=1}^{P} r^2(\tau). \tag{2.3}$$

Since the $r(\tau)$'s are independently distributed in large samples, Q will be asymptotically χ_P^2 for data from a white noise process. If a number of the sample autocorrelations are not close to zero, Q will be inflated.

The choice of P is somewhat arbitrary. Setting P high means that we capture what might be highly significant $r(\tau)$'s at relatively high lags. On the other hand, by making P high, a good deal of power may be lost. As an example, suppose that the true model is an MA(1) process. While a test based on $r(1)$ or the von Neumann Ratio may have a relatively high power against such an alternative, the portmanteau test may be ineffective. Although $r(1)$ may be some way from zero, its influence in the Q-statistic will be diluted by the remaining $P-1$ autocorrelations. The portmanteau test should not therefore be regarded as a substitute for examining the correlogram, since an insignificant result may easily mask some distinctive feature.

The justification for the portmanteau test is an asymptotic one. However, in small samples it has been found that χ_P^2 does not provide a particularly good approximation to the distribution of Q under the null hypothesis. A modification, which is rather more satisfactory in this respect, is

$$Q^* = T(T+2) \sum_{\tau=1}^{P} (T-\tau)^{-1} r^2(\tau). \tag{2.4}$$

The rationalisation for (2.4) stems from the recognition that the expression $(T-\tau)/T(T+2)$ provides a closer approximation to the variance of $r(\tau)$ than does $1/T$; see Ljung and Box (1978).

The justification for the portmanteau test has so far been on pragmatic grounds only. Any departures from randomness will, almost by definition, be reflected in the sample autocorrelations, and the portmanteau statistic is constructed in such a way as to pick this up. Furthermore, the form of its asymptotic distribution is immediately apparent given the distribution of the sample auto-correlations. A more formal rationale for the portmanteau test may, however, be given: if the null hypothesis of a white noise process is tested against the alternative of an $AR(P)$ or an $MA(P)$ model, then the Lagrange multiplier test is identical to a portmanteau test based on (2.3). To the extent that a high order AR or MA process may be regarded as an approximation to any linear process, the portmanteau test is a general test against non-randomness.

Consider the alternative of an $AR(P)$ model and assume, for simplicity, that the process is known to have zero mean. Since the asymptotic information matrix for an $AR(P)$ process is block diagonal with respect to ϕ and σ^2, the LM statistic is given directly by (1.3.5) with $\psi = \phi$. From (5.4.2),

$$IA(\phi) = V_P, \tag{2.5}$$

where $\sigma^2 V_P$ is the covariance matrix of the observations over P successive time periods. Under the null hypothesis, $H_0 : \phi_1 = \phi_2 = \cdots = \phi_P = 0$, V_P reduces to the identity matrix, and so the LM statistic is

$$LM = T^{-1} \left(\frac{\partial \log L}{\partial \phi} \right)' \left(\frac{\partial \log L}{\partial \phi} \right) = T^{-1} \sum_{i=1}^{P} \left(\frac{\partial \log L}{\partial \phi_i} \right)^2 \tag{2.6}$$

evaluated at $\phi = 0$. The conditional log–likelihood function is

$$\log L = -(1/2)T \log 2\pi - (1/2)T \log \sigma^2 - (1/2)\sigma^{-2} S(\phi),$$

where $S(\phi)$ is defined in (5.2.3). Differentiating with respect to the elements of ϕ gives

$$\frac{\partial \log L}{\partial \phi_i} = \sigma^{-2} \sum_{t=P+1}^{T} y_{t-i}(y_t - \phi_1 y_{t-1} - \cdots - \phi_P y_{t-P}),$$

$$i = 1, \ldots, P. \tag{2.7}$$

Under H_0, $\tilde{\sigma}_0^2 = T^{-1} \sum y_t^2$, and (2.7) becomes

$$\left. \frac{\partial \log L}{\partial \phi_i} \right|_{\phi=0} = \tilde{\sigma}_0^{-2} \sum_{t=P+1}^{T} y_{t-i} y_t = Tr(i), \qquad i = 1, \ldots, P. \tag{2.8}$$

On substituting in (2.6) it follows immediately that

$$LM = T \sum_{\tau=1}^{P} r^2(\tau).$$

This is identical to the portmanteau test statistic, (2.3), except that $r(\tau)$ is defined without $\bar{y}$ appearing in the autocovariances. This is simply a consequence of the assumption that the process has zero mean.

The portmanteau test may also be derived as a test against an MA(P) alternative. As with the AR(P) process, the information matrix under H_0 is equal to T multiplied by a $P \times P$ identity matrix. The LM statistic is similar in form to (2.6), but with $\partial \log L / \partial \phi_i$ replaced by

$$\frac{\partial \log L}{\partial \theta_j} = \sigma^{-2} \sum_{t=1}^{T} \left(\epsilon_{t-j} + \sum_{i=1}^{q} \theta_i \frac{\partial \epsilon_{t-i}}{\partial \theta_j} \right) \epsilon_t, \qquad j = 1, \ldots, P.$$

Both the residual, ϵ_t, and its derivatives obey recursions of the form (5.3.3) and (5.3.4), but under the null hypothesis $\theta = 0$ and $\epsilon_t = y_t$. Thus

$$\left. \frac{\partial \log L}{\partial \theta_j} \right|_{\theta=0} = Tr(j), \qquad j = 1, \ldots, P, \tag{2.9}$$

and the argument goes through as before.

An alternative approach in both the AR and MA cases is to equate maximising the likelihood function with minimising the conditional sum of squares. A test may then be based on the TR^2 statistic, where R^2 is the coefficient of multiple correlation obtained by regressing ϵ_t on z_t, its vector of partial derivatives. In both cases, this amounts simply to regressing y_t on $y_{t-1}, y_{t-2}, \ldots, y_{t-P}$. The TR^2 statistic is then tested as a χ^2 variate with P degrees of freedom.

Cumulative Periodogram

The cumulative periodogram is a frequency domain alternative to the portmanteau test. It is based on the periodogram ordinates, p_j^2, defined in Section 3.2. These are used to construct a series of statistics,

$$s_i = \sum_{j=1}^{i} p_j^2 \bigg/ \sum_{j=1}^{n} p_j^2, \qquad i = 1, \ldots, n. \tag{2.10}$$

The test procedure is based on a plot of s_i against i, which is known as the cumulative periodogram. This differs from the periodogram itself, in that the highly unstable behaviour of the p_j's is, to a large extent, ironed out by the process of accumulation. Thus, although a visual inspection of the periodogram is of limited value, the cumulative periodogram is a useful diagnostic tool.

For a white noise series, the s_i's will lie close to the 45° line on the graph of s_i against i. On the other hand, the cumulative periodogram

for a process with an excess of low frequency will tend to lie above
the 45° line; cf. figure 3.2. By way of contrast, a process with an
excess of high frequency components will tend to have a cumulative
periodogram which lies below the 45° line.

A formal test for departures from randomness is obtained by
constructing two lines parallel to the 45° line, $s = i/n$. These are
defined by

$$s = \pm c_0 + i/n, \tag{2.11}$$

where c_0 is a significance value which depends on n and may be read
off directly from a table given in Durbin (1969). This table is
reproduced in EATS as table C (pp. 364–5), and the appropriate
significance value is obtained by entering the table at $n - 1$. For a
two-sided test of size α, c_0 is read from the column headed $\alpha/2$. The
null hypothesis is rejected at the α level of significance if the sample
path, $s_1, \ldots, s_n$, crosses either of the lines in (2.11).

In certain circumstances a one-sided test may be appropriate. For
example, if the alternative hypothesis is that there is an excess of low
frequency, only the upper line is relevant. Since an excess of low
frequency corresponds to positive serial correlation, such an
alternative will often be very reasonable. The significance value is
found in exactly the same way as for a one-sided test, except that
the column headed 'α' is now the one to be consulted.

When T is odd, the one-sided test is exact. However, the
approximations involved in carrying out two-sided tests, or one-sided
tests with T even, are likely to be negligible in practice. Note that
the test can be carried out without actually graphing the s_i's. The
rule for a two-sided procedure is to reject H_0 if

$$\max_i | s_i - i/n | > c_0.$$

3. Model Selection

The previous chapter discussed the estimation of the parameters in
an ARMA(p, q) model for *given* values of p and q. In modelling a
particular stationary time series, however, p and q will not, in general,
be known. Prior to estimation, therefore, a suitable model must be
selected. This involves choosing p and q in such a way that a good fit
is obtained with a minimum number of parameters. Box and Jenkins
refer to this as the principle of *parsimony*.

In Section 5.8 of EATS, certain general principles for model
selection were laid out. It was argued that the optimal approach is to
start from the most general model and to successively test more and

more stringent restrictions. This approach would be feasible in the present context if it were reasonable to regard a pure AR (or pure MA) process as the most general formulation. Thus, given an AR(p) model, the first hypothesis to be tested is that the process is of order p. The next hypothesis in the nest is that the process is of order $p - 1$ and so on. The order of the model is then fixed when a particular hypothesis is rejected. Full details of such an approach will be found in Anderson (1971, pp. 270–6). Unfortunately, the procedure is open to two objections, one minor and the other major. The minor objection is that the set of ordered hypotheses which have been defined may not always be appropriate. The best model may have an order greater than the number of non-zero parameters; for example, an AR(2) model may have $\phi_1 = 0$. However, except in seasonal cases, such formulations are relatively uncommon, and once p has been determined it would not be difficult to detect zero coefficients in any case. The more basic objection is that it may take a relatively large number of AR parameters to approximate a particular linear process at all well. This has two implications. Firstly, the initial general model should be chosen with a sufficiently high order. Secondly, once a model has been selected by the sequential testing procedure it may well contain a very large number of parameters, thereby violating the principle of parsimony.

A mixed ARMA model has the advantage that it can usually approximate a linear process with relatively few parameters. The problem with a mixed model is that there is no longer any natural ordering of hypotheses. Furthermore, a technical difficulty arises when such a model is overparameterised. As a simple example suppose that an ARMA(1, 1) model

$$y_t = \phi y_{t-1} + \epsilon_t + \theta \epsilon_{t-1}, \tag{3.1}$$

is fitted to a set of observations from a white noise process. Model (3.1) reduces to white noise when $\phi = -\theta$, but when this constraint is imposed the asymptotic information matrix is singular; cf. (5.4.7). This is likely to cause computational problems, since most ML estimation procedures rely on a matrix which is asymptotically equivalent to the information matrix. If the routine does not actually fail, but converges to estimates for which $\tilde{\phi} \neq -\tilde{\theta}$, an important clue to the over-parameterisation will lie in the estimated standard errors, which because of the near singularity of the estimated information matrix will tend to be high.

The above considerations led Box and Jenkins to advocate a strategy for model selection based on a three stage iterative procedure. In the first stage, *identification*, a tentative model is selected on the basis of the appearance of the correlogram and other related statistics. Given p and q, the parameters are then estimated by one of the methods described in the previous chapter. As a by-product of this *estimation* stage, residuals are computed and these may be used as the basis for *diagnostic checking*. A significant departure from randomness indicates that the model is, in some sense, inadequate. A return to the identification stage is then called for, and the complete cycle is repeated until a suitable formulation is found.

Identification

The correlogram provides the most important means by which a suitable ARMA(p, q) process may be identified. Provided the sample is reasonably large, the correlogram should mirror the theoretical autocorrelation function of the underlying process. In Chapter 2 the behaviour of the autocorrelation function was examined and certain salient characteristics were noted for particular processes. Thus, a pure MA(q) process exhibits a cut-off in its autocorrelation function for lags greater than q. On the other hand, the autocorrelations of an AR or mixed process gradually tail off to zero. A correlogram of the type shown in figure 6.1 would therefore probably lead to an MA(2) model being initially entertained. While the autocorrelations beyond $\tau = 2$ are not identically equal to zero, they are small enough in comparison with $r(1)$ and $r(2)$ to suggest that their size is primarily due to sampling error.

If the theoretical autocorrelations beyond a certain point, q, are zero, it can be shown that the sample autocorrelations are approximately normally distributed in large samples with mean zero and variance

$$\text{Avar}[r(\tau)] = \left[1 + 2 \sum_{j=1}^{q} \rho(j)\right] \Big/ T, \qquad \tau > q. \tag{3.2}$$

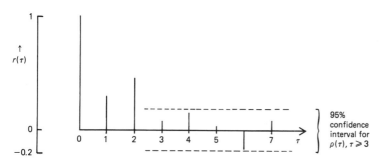

Figure 6.1   *Sample Autocorrelations for 200 Observations from an MA(2) Process*

An estimate of Avar$[r(\tau)]$ can be obtained by replacing the theoretical autocorrelations in (3.2) by the corresponding sample autocorrelations. The broken lines in figure 6.1 indicate approximate 95% confidence intervals for each $r(\tau)$, $\tau > 2$, under the assumption that the process is indeed MA(2).

The order of a pure AR process is rather more difficult to determine from the correlogram, except perhaps when $p = 1$. A complementary procedure is therefore often used. This is based on the sample *partial autocorrelation function*. Let $\tilde{\phi}(\tau)$ denote the estimated coefficient of $y_{t-\tau}$ in an AR(τ) model. The sample partial autocorrelation function is then defined by a plot of $\tilde{\phi}(\tau)$ against τ. The important point to note about this function is that its behaviour is the opposite of that exhibited by the autocorrelation function. For a pure AR(p) process, the theoretical partial autocorrelations are zero at lags beyond p, while for an MA process they die away gradually. If the observations are generated by an AR(p) process, the sample partial autocorrelations beyond lag p are normally distributed with mean zero and variance,

$$\text{Avar}[\tilde{\phi}(\tau)] = 1/T. \tag{3.3}$$

If the data can be reasonably well approximated by a pure AR or MA model, it should not prove too difficult to select a suitable value of p or q by examining the estimated autocorrelation, or partial autocorrelation, function. The identification of both p and q in a mixed model is somewhat more problematic. Neither the autocorrelation function nor the partial autocorrelation function has a definite cut-off point, and considerable skill may be needed to interpret the patterns obtained. In these circumstances the complete cycle of identification, estimation and diagnostic checking may have to be repeated several times before a suitable model is found.

The above difficulties may be seriously compounded by sampling error. For an underlying AR(1) process, Kendall (1973, Chapter 7) notes that the correlation between $r(\tau)$ and $r(\tau + k)$ is approximately

$$\left(\frac{1}{T}\right) \frac{\phi^k[(k+1) - (k-1)\phi^2]}{1 - \phi^2}. \tag{3.4}$$

Thus with ϕ taking the fairly moderate value of 0.5, the correlation between successive values of $r(\tau)$ is 0.80. The effect of the sampling error is therefore likely to be cumulative, with the result that the correlogram may provide a distorted picture of the underlying autocorrelation function. A further related consequence is that the correlogram may not damp down as fast as the corresponding autocorrelation function. Ripples and trends which have no basis in the theoretical function can easily appear in the estimated function. The graphs in Kendall (1973, pp. 84 and 90) provide an excellent illustration of this phenomenon.

Diagnostic Checking

Diagnostic checking of the model may be carried out by examining the residuals for departures from randomness. Formal tests can be based on the correlogram and the cumulative periodogram, although neither of these should be seen as a substitute for a direct plot of the residuals. To quote Box and Jenkins (1976, p. 289): 'It cannot be too strongly emphasized that visual inspection of a plot of the residuals themselves is an indispensable first step in the checking process.'

Although plotting and examining functions of the residuals is an extremely valuable exercise, some care should be taken in interpreting the results. Tests associated with these procedures are often constructed on the assumption that the residuals have the same properties as the disturbances when the model is correctly specified. Unfortunately this assumption is invalid for residuals from ARMA models, even when the sample size is large. For example, if an AR(1) model is fitted to the data, it can be shown that the first-order autocorrelation in the residuals has an asymptotic variance of ϕ^2/T. This can be substantially less than $1/T$, which is the variance of $r(1)$ for a white noise process. However, for higher order lags the bias in variance is considerably less serious. This is indicative of the behaviour of the residual autocorrelations for any fitted $\text{ARMA}(p, q)$ process. In all cases, a reduction in variance tends to occur at low lags. Furthermore, the $r(\tau)$'s at these lags can be highly correlated. Although such effects usually disappear at higher lags, they are often sufficiently important to impart a severe bias to the portmanteau statistic and the cumulative periodogram. The net result is that these test procedures tend to underestimate the significance of apparent discrepancies.

Although the tests associated with the graphical procedures are not generally valid in this situation, they can nevertheless provide useful guidelines. However, a valid test statistic is clearly desirable. Box and Pierce (1970) have shown that, provided P is reasonably large, the portmanteau statistic constructed from ARMA residuals is appropriate if it is taken to have a χ^2_{P-p-q} distribution under the null hypothesis. The reduction in degrees of freedom is to allow for the fitted parameters, and the net result is that the probability of rejecting the null hypothesis is increased as compared with a portmanteau test based on a χ^2 distribution with the full P degrees of freedom. Unfortunately the statistic still suffers from the small sample defects described in Section 2, and the preferred approach is now to calculate the modified portmanteau statistic (2.4). This is then tested against an appropriate significance value from the χ^2_{P-p-q} distribution. When (2.3) and (2.4) are used in this way, the resulting procedures are often referred to as the *Box—Pierce test* and the *modified Box— Pierce* (or Box—Ljung) *test* respectively.

An alternative approach to testing the specification of an ARMA(p, q) model is to nest it within an ARMA$(p + P, q)$ process. A Lagrange multiplier test of the hypothesis $H_0 : \phi_{p+1} = \phi_{p+2} = \cdots = \phi_{p+P} = 0$ is then carried out. The residual ϵ_t is regressed on the full set of $p + P + q$ derivatives evaluated under H_0 and the resulting TR^2 statistic is treated as a χ^2_P variate. The 'modified' LM procedure consists of carrying out an asymptotic F-test on the joint significance of the derivatives with respect to ϕ_{p+1} to ϕ_{p+P}.

Example 1   Suppose that an ARMA$(1, 1)$ model has been estimated. A test against an ARMA$(1 + P, 1)$ specification involves the evaluation of the derivatives of the residual function,

$$\epsilon_t = y_t - \phi_1 y_{t-1} - \cdots - \phi_{1+P} y_{t-1-P} - \theta \epsilon_{t-1},$$

$$t = 2 + P, \ldots, T \quad (3.5)$$

with $\epsilon_{1+P} = 0$. These derivatives are

$$\frac{\partial \epsilon_t}{\partial \phi_i} = -y_{t-i} - \theta \frac{\partial \epsilon_{t-1}}{\partial \phi_i} = - \sum_{j=0}^{t-P-2} (-\theta)^j y_{t-i-j},$$

$$t = 2 + P, \ldots, T, \quad (3.6a)$$

for $i = 1, \ldots, 1 + P$, and

$$\frac{\partial \epsilon_t}{\partial \theta} = -\epsilon_{t-1} - \theta \frac{\partial \epsilon_{t-1}}{\partial \theta} = - \sum_{j=0}^{t-P-2} (-\theta)^j \epsilon_{t-1-j},$$

$$t = 2 + P, \ldots, T. \quad (3.6b)$$

Note that since ϵ_{1+P} is assumed to be fixed, all the derivatives at $t = 1 + P$ are equal to zero.

Evaluating these expressions under the null hypothesis leads to a regression of $y_t - \phi_1 y_{t-1} - \theta \epsilon_{t-1}$ on (3.6) with ϕ_1, θ and the ϵ_t's replaced by their estimated values. The resulting TR^2 statistic is tested against a χ^2_P distribution.

An LM test could also be set up by augmenting the MA component to yield an ARMA$(p, q + P)$ process. (Example 5.3 is a special case.) At first sight this would seem to yield a different test to the one described above, but in fact it turns out that the two TR^2 statistics are identical; see Poskitt and Tremayne (1980).

A final point to note about these LM tests is that they are valid for any value of P. This is an advantage over the Box–Pierce test, where a reasonably large value of P, say $P = \log T$, is needed for the asymptotic theory to go through.

Goodness of Fit

In a given application, the Box–Jenkins model selection procedure may suggest several specifications, each of which satisfies the diagnostic checks. Some kind of measure of goodness of fit is therefore needed to distinguish between different models in these circumstances. One possibility is the Akaike Information Criterion, described in Section 1.3. This can be conveniently expressed in the form

$$\text{AIC}^\dagger = \bar{\sigma}^2 \exp[\log|V| + 2(p+q)/T], \tag{3.7}$$

where $\sigma^2 V$ is the covariance matrix of y. If the model is estimated by CSS rather than full ML, the term $\log|V|$ disappears.

The preferred model is the one for which $\text{AIC}^\dagger$ is a minimum. The term $2(p+q)/T$ assigns a penalty to models which are not suitably parsimonious.

One of the disadvantages of the Box–Jenkins strategy is that it requires a good deal of time and skill, particularly at the identification stage. An alternative approach is to use the AIC more comprehensively. The range of possible models is defined on the basis of a priori information, or by a rather crude identification procedure, and the one for which (3.7) is a minimum is selected. This implies a shift in emphasis towards the computer. However, although the procedure may be applied in a semi-automatic fashion, it would be unwise not to submit a model chosen in this way to diagnostic checking.

4. Prediction

The concept of a minimum mean square estimator (MMSE) was introduced in Chapter 4. Such an estimator minimises the variance of the prediction error, and so if $\tilde{y}_{T+l/T}$ is the MMSE of y_{T+l}, the quantity $E[(y_{T+l} - \tilde{y}_{T+l/T})^2]$ will be equal to or less than the MSE for all other predictors.

Three assumptions will be made initially. The first is that the disturbances term, ϵ_t, is normally distributed. This enables the discussion to be restricted to linear estimators, since under normality the MMSE is identical to the MMSLE. The second assumption is that the ARMA parameters are *known*. Although this will rarely be the case in practice, it can be shown that estimation errors in the parameters will not seriously affect the predictions unless the sample size is small. This point is examined in some detail in the last sub-section. Finally, for MA and mixed processes it will be assumed that all present and past disturbances, i.e. $\epsilon_T, \epsilon_{T-1}, \ldots$, are known. This is effectively the same as assuming an infinite realisation of observations. The assumption is not a restrictive one unless T is small and/or some of the roots of the MA polynomial lie close to or on the unit circle. When such considerations are important, *finite sample prediction* is appropriate and a method of effecting this is described in a later sub-section.

The equation of an ARMA(p, q) process at time $T + l$ is given by

$$y_{T+l} = \phi_1 y_{T+l-1} + \cdots + \phi_p y_{T+l-p} + \epsilon_{T+l} + \cdots + \theta_q \epsilon_{T+l-q}.$$

$$(4.1)$$

Future values of ϵ_t are unknown, and cannot be predicted. Thus if (4.1) is used as the basis for predicting y_{T+l}, the ϵ_t's must be set equal to their expected values of zero for $t > T$. Future values of y_t are also unknown, but predictions may be computed from the recursion

$$\tilde{y}_{T+l/T} = \phi_1 \tilde{y}_{T+l-1/T} + \cdots + \phi_p \tilde{y}_{T+l-p/T}$$
$$+ \tilde{\epsilon}_{T+l/T} + \cdots + \theta_q \tilde{\epsilon}_{T+l-q/T}, \qquad l = 1, 2, \ldots, \quad (4.2)$$

where $\tilde{y}_{T+j/T} = y_{T+j}$ for $j \leqslant 0$, and

$$\tilde{\epsilon}_{T+j/T} = \begin{cases} 0 & \text{for } j > 0 \\ \epsilon_{T+j} & \text{for } j \leqslant 0. \end{cases}$$

Example 1 For the AR(1) process, expression (4.2) leads directly to the difference equation

$$\tilde{y}_{T+l/T} = \phi \tilde{y}_{T+l-1/T}, \qquad l = 1, 2, \ldots \qquad (4.3)$$

The starting value is given by $\tilde{y}_{T/T} = y_T$, and so (4.3) may be solved to yield

$$\tilde{y}_{T+l/T} = \phi^l y_T. \qquad (4.4)$$

Thus the predicted values decline exponentially towards zero, and the forecast function has exactly the same form as the auto-covariance function.

Example 2   At time $T + 1$, the equation for an MA(1) process is of the form

$$y_{T+1} = \epsilon_{T+1} + \theta\epsilon_T. \tag{4.5}$$

Since ϵ_{T+1} is unknown, it is set equal to zero in the corresponding prediction equation which is

$$\tilde{y}_{T+1/T} = \theta\epsilon_T. \tag{4.6}$$

For $l > 1$, $\tilde{y}_{T+l/T} = 0$, and so knowledge of the data generating process is of no help in predicting more than one period ahead.

Example 3   Consider the ARMA(2, 2) process

$$y_t = 0.6y_{t-1} + 0.2y_{t-2} + \epsilon_t + 0.3\epsilon_{t-1} - 0.4\epsilon_{t-2} \tag{4.7}$$

and suppose that $y_T = 4.0$, $y_{T-1} = 5.0$, $\epsilon_T = 1.0$ and $\epsilon_{T-1} = 0.5$. Then

$$\tilde{y}_{T+1/T} = 0.6y_T + 0.2y_{T-1} + 0.3\epsilon_T - 0.4\epsilon_{T-1} = 3.5$$

and

$$\tilde{y}_{T+2/T} = 0.6\tilde{y}_{T+1/T} + 0.2y_T - 0.4\epsilon_T = 2.5.$$

Thereafter forecasts are generated by the difference equation

$$\tilde{y}_{T+l/T} = 0.6\tilde{y}_{T+l-1/T} + 0.2\tilde{y}_{T+l-2/T}, \qquad l = 3, 4, \ldots$$

The Optimal Predictor and its MSE

In the case of the MA(1) model of Example 2, the MSE of $\tilde{y}_{T+1/T}$ may be obtained immediately. From (4.5) and (4.6)

$$\text{MSE}(\tilde{y}_{T+1/T}) = E[(y_{T+1} - \tilde{y}_{T+1/T})^2] = E(\epsilon_{T+1}^2) = \sigma^2.$$

For $l > 1$ the solution is trivial, since $\tilde{y}_{T+l/T} = 0$ and $\text{MSE}(\tilde{y}_{T+l/T}) = \text{Var}(y_{T+l}) = (1 + \theta^2)\sigma^2$.

For any MA(q) process it is not difficult to see that the calculation of the prediction MSE is straightforward. The solution for AR and mixed processes is not immediately apparent. However, any stationary ARMA process may be expressed as an infinite moving average,

$$y_t = \sum_{j=0}^{\infty} \psi_j \epsilon_{t-j}, \tag{4.8}$$

where the ψ_j's may be determined from (2.4.6). Since any predictor linear in the observations must also be linear in the disturbances, it must be of the form

$$\hat{y}_{T+l/T} = \psi_l^* \epsilon_T + \psi_{l+1}^* \epsilon_{T-1} + \cdots, \qquad l = 1, 2, \ldots, \qquad (4.9)$$

as future values of ϵ_t are unknown. The coefficients $\psi_l^*, \psi_{l+1}^*, \ldots$ are weights. Since

$$y_{T+l} - \hat{y}_{T+l/T} = \epsilon_{T+l} + \psi_1 \epsilon_{T+l-1} + \cdots + \psi_{l-1} \epsilon_{T+1}$$
$$+ (\psi_l - \psi_l^*)\epsilon_T + (\psi_{l+1} - \psi_{l+1}^*)\epsilon_{T-1} + \cdots$$

has zero expectation, $\hat{y}_{T+l/T}$ is u-unbiased, and its MSE is given by

$$\text{MSE}(\hat{y}_{T+l/T}) = \sigma^2(1 + \psi_1^2 + \cdots + \psi_{l-1}^2) + \sigma^2 \sum_{j=0}^{\infty} (\psi_{l+j} - \psi_{l+j}^*)^2.$$

$$(4.10)$$

This is minimised by setting $\psi_{l+j}^* = \psi_{l+j}$. The MMSLE of y_{T+l} is therefore

$$\tilde{y}_{T+l/T} = \sum_{j=0}^{\infty} \psi_{l+j} \epsilon_{T-j} \qquad (4.11)$$

and

$$\text{MSE}(\tilde{y}_{T+l/T}) = (1 + \psi_1^2 + \cdots + \psi_{l-1}^2)\sigma^2. \qquad (4.12)$$

The predictions given by (4.2) are identical to those obtained from (4.11).

Example 4   For the AR(1) model $\psi_{l+j} = \phi^{l+j}$, and so

$$\tilde{y}_{T+l/T} = \sum_{j=0}^{\infty} \phi^{l+j} \epsilon_{T-j} = \phi^l \sum_{j=0}^{\infty} \phi^j \epsilon_{T-j} = \phi^l y_T.$$

cf. (4.4). The MSE of $y_{T+l/T}$ in the AR(1) model is given by

$$\text{MSE}(\tilde{y}_{T+l/T}) = [1 + \phi^2 + \cdots + \phi^{2(l-1)}]\sigma^2$$

$$= \frac{1 - \phi^{2l}}{1 - \phi^2} \cdot \sigma^2. \qquad (4.13)$$

As $l \to \infty$, this expression tends towards $\sigma^2/(1 - \phi^2)$ which is simply the variance of y_t.

Since the MMSE of y_{T+l} is a linear function of the disturbances, it will be normally distributed when the disturbances are normal. A 95% prediction interval is given by

$$y_{T+l} = \tilde{y}_{T+l/T} \pm 1.96 \left(1 + \sum_{j=1}^{l-1} \psi_j^2 \right)^{1/2} \sigma. \qquad (4.14)$$

Finite Sample Prediction and Invertibility*

In the MA(1) model, the MMSE of y_{T+1} depends on ϵ_T. However, even though θ may be known, ϵ_T can never be determined exactly from a *finite* sample of observations $y_1, \ldots, y_T$. In practice, the prediction will usually be carried out with ϵ_T replaced by $e_T = \epsilon_T(\theta)$, the residual obtained from the CSS recursion, (5.3.3), with starting value $\epsilon_0(\theta) = 0$.

The consequence of replacing ϵ_T by e_T is an increase in the MSE of the predictor. Repeated substitution for lagged values of ϵ_t in the expression $\epsilon_t = y_t - \theta \epsilon_{t-1}$ yields

$$\epsilon_t = y_t + (-\theta)y_{t-1} + \theta^2 y_{t-2} + \cdots + (-\theta)^{t-1}y_1 + (-\theta)^t \epsilon_0.$$

This corresponds to the expression for e_t given by the CSS recursion, except that ϵ_0 is replaced by zero. Substituting in the MSE formula therefore gives

$$\mathrm{MSE}(\hat{y}_{T+1/T}) = E\{[\epsilon_{T+1} + \theta(\epsilon_T - e_T)]^2\} = \sigma^2[1 + \theta^{2(T+1)}].$$

$$(4.15)$$

If T is large, the difference between ϵ_T and e_T will be negligible, provided that $|\theta| < 1$. This is reflected in formula (4.15) which tends to σ^2 as $T \to \infty$. On the other hand, if T is relatively small and $|\theta|$ is close to unity, the actual predictor may be significantly different from (4.6). The most extreme case arises if θ lies on the unit circle. The process is then strictly non-invertible and since the prediction error is given by $y_{T+1} - \theta e_T = \epsilon_{T+1} - (-\theta)^{T+1}\epsilon_0 = \epsilon_{T+1} \pm \epsilon_0$, it remains dependent on ϵ_0, irrespective of the sample size. The MSE of the predictor is equal to $2\sigma^2$.

The problem of obtaining the exact MMSE of a future observation for an MA or mixed process is sometimes referred to as the *finite sample prediction* problem. One solution is by means of the Kalman filter. This was described in Section 5.3 in the context of computing the likelihood function. For given values of ϕ and θ, the MMSE of y_{T+l} may be obtained by using the results on multi-step prediction set out in Section 4.5. In the MA(1) case $\mathrm{MSE}(\tilde{y}_{T+1/T}) = \sigma^2 f_{T+1}$, where f_{T+1} is defined by (4.4.9). A comparison with (4.15) shows this to be less than the MSE of the quasi-MMSE, except when $\theta = 0$. When $\theta = \pm 1$, result (4.5.15) ensures that $\mathrm{MSE}(\tilde{y}_{T+1/T}) \to \sigma^2$ as $T \to \infty$.

As a final point note that finite sample predictions can be made even if the process is non-invertible in the general sense. In the case of an MA(1) process, this means $|\theta| > 1$. The reason why forecasting from such a model is unattractive is that it is inefficient. Using (5.3.9) it can be seen that $\mathrm{MSE}(\tilde{y}_{T+1/T}) \to \sigma^2 \theta^2$ when $|\theta| > 1$, whereas the corresponding invertible model has $\text{MSE}(\tilde{y}_{T+1/T}) \to \sigma^2$ as $T \to \infty$.

Prediction with Estimated Parameters*

In practice, predictions are almost invariably made with ϕ and θ replaced by their estimates. This creates an additional source of variability, which should ideally be incorporated in the expression for the prediction MSE.

Consider the AR(1) process. When ϕ is known, the MMSE for l periods ahead is given by (4.4). When ϕ is unknown it will be replaced by its ML estimator, $\tilde{\phi}$, or by an estimator which is asymptotically equivalent. The actual predictor is therefore

$$\tilde{y}^*_{T+l/T} = \tilde{\phi}^l y_T. \tag{4.16}$$

In more general cases, $\tilde{y}^*_{T+l/T}$ can be computed by a difference equation having exactly the same form as (4.2).

The prediction error for (4.16) may be decomposed into two parts by writing

$$y_{T+l} - \tilde{y}^*_{T+l/T} = (y_{T+l} - \tilde{y}_{T+l/T}) + (\tilde{y}_{T+l/T} - \tilde{y}^*_{T+l/T}). \tag{4.17}$$

The first term on the right-hand side of (4.17) is the prediction error when ϕ is known, while the second term represents the error arising from the estimation of ϕ. This decomposition is appropriate for all ARMA models. In the present case it specialises to

$$y_{T+l} - \tilde{y}^*_{T+l/T} = (y_{T+l} - \tilde{y}_{T+l/T}) + (\phi^l - \tilde{\phi}^l)y_T \tag{4.18}$$

in view of (4.16).

Now consider the one-step ahead predictor. The MSE may be written as

$$\text{MSE}(\tilde{y}^*_{T+1/T}) = \text{MSE}(\tilde{y}_{T+1/T}) + y_T^2 E[(\tilde{\phi} - \phi)^2]. \tag{4.19}$$

In formulating the contribution of the estimation error to (4.19), y_T is taken as fixed, whereas $\tilde{\phi}$ is a random variable. This may appear to be contradictory, as y_T is actually used to construct $\tilde{\phi}$. However, (4.19) provides a sensible definition of MSE in this context, since any prediction is always made conditional on the sample observations being known. Replacing $E[(\tilde{\phi} - \phi)^2]$ by its asymptotic variance gives an approximation to the mean square error, i.e.

$$\text{MSE}(y^*_{T+l/T}) \simeq \sigma^2 + y_T^2(1 - \phi^2)/T. \tag{4.20}$$

The AR(1) model is often estimated by OLS. Applying the usual regression formula for estimating the MSE of $\tilde{y}^*_{T+1/T}$ gives

$$\text{mse}(y_{T+1/T}^*) = s^2 \left(1 + y_T^2 \middle/ \sum_{t=2}^{T} y_{t-1}^2\right).$$ (4.21)

This is closely related to (4.20) since

$$\sum_{t=2}^{T} y_{t-1}^2 \simeq T\sigma^2/(1 - \phi^2)$$

in large samples. Fuller and Hasza (1978) have recently provided evidence indicating that the usual regression formula is, in fact, an excellent estimator of the prediction MSE for multi-step prediction in more general AR models.

When $l > 1$, the last term in (4.19) is $y_T^2 E[(\tilde{\phi}^l - \phi^l)^2]$. Writing

$$\phi^l - \tilde{\phi}^l = \phi^l - \phi^l\left[1 - \frac{(\phi - \tilde{\phi})}{\phi}\right]^l$$

and expanding the term in square brackets yields

$$\phi^l - \tilde{\phi}^l \simeq l\phi^{l-1}(\phi - \tilde{\phi})$$

when higher order terms are ignored. Therefore

$$E[(\tilde{\phi}^l - \phi^l)^2] \simeq l^2\phi^{2(l-1)}E[(\tilde{\phi} - \phi)^2].$$

Together with the result in (4.13) this gives

$$\text{MSE}(y_{T+l/T}^*) \simeq \sigma^2 \frac{1 - \phi^{2l}}{1 - \phi^2} + y_T^2 \frac{(1 - \phi^2)l^2\phi^{2(l-1)}}{T}.$$ (4.22)

Expression (4.22) is an approximation to the MSE of the multi-step predictor for a particular sample. In order to get some idea of the MSE of such a predictor *on average*, y_T^2 is replaced by its expected value $\sigma^2/(1 - \phi^2)$. The resulting expression is known as the *asymptotic mean square error* (AMSE). Thus

$$\text{AMSE}(y_{T+l/T}^*) \simeq \sigma^2\left[\frac{1 - \phi^{2l}}{1 - \phi^2} + \frac{l^2\phi^{2(l-1)}}{T}\right], \qquad l = 1, 2, \dots$$ (4.23)

For the special case of $l = 1$,

$$\text{AMSE}(y_{T+1/T}^*) = \sigma^2(1 + T^{-1}).$$ (4.24)

In both (4.23) and (4.24), the contribution arising from the error in estimating ϕ is a term of $0(T^{-1})$. This will be dominated by the expression for the MSE of the optimal predictor when ϕ is known. Although ignoring the effect of estimating ϕ will lead to an under-

estimate of the variability of the prediction error, the bias is unlikely
to be severe unless T is very small. These findings carry over to more
general models. Further discussion will be found in Box and
Jenkins (1976, pp. 267–9) and Yamamoto (1976).

5. Non-Stationarity

The time series models developed up to this point have rested
crucially on the assumption of stationarity. In many practical
situations this assumption is clearly too restrictive. A plot of a time
series will frequently show some kind of trend in the mean and
possibly in the variance also.

There are basically two ways in which the ARMA framework may
be extended to handle non-stationarity. The first is by attempting to
construct a stationary series by differencing, and then to model this
differenced series as an ARMA process. Usually only first or second
differences are required, and the full model is referred to as an
integrated process. The second approach is to postulate a *trend plus
error* model of the form

$$y_t = f(t) + u_t, \tag{5.1}$$

where $u_t \sim \text{ARMA}(p, q)$ and $f(t)$ is a deterministic function of time.

In the absence of any prior knowledge on the form of the
deterministic trend in (5.1), the obvious thing to do is to model it as
a polynomial. Thus

$$f(t) = \alpha + \beta_0 t + \beta_1 t^2 + \cdots + \beta_h t^h, \tag{5.2}$$

where h is the order of the polynomial. The disadvantage of this
approach, or indeed of any approach based on (5.1), is that it assumes
a basic underlying trend throughout the length of the series. It is
therefore difficult to accommodate any changes in direction towards
the end of the series and the forecasting performance of the model
may be poor. Differencing offers greater flexibility in this respect,
since it replaces the *global* trend of (5.2) by a *local* trend. Forecasts
are still constructed by extrapolating polynomials, but the later
observations now receive a much greater weight in determining the
starting values; cf. the discussion in Section 1.1.

The first sub-section below examines the properties of integrated
processes, while the second sub-section describes the mechanics of
making predictions. Formulae for determining the mean square error
of forecasts are given. Estimating the MSE is crucial to any scientific
attempt at forecasting, since it provides an explicit statement on the
reliability of predictions at any lead time. The final sub-section

provides something of a link between local and global trend models, in that it examines the consequences of replacing the deterministic function in (5.1) by a stochastic trend.

There are an infinite number of ways in which non-stationary observations can be generated. A trending mean is only one aspect of this phenomenon. No attempt will be made here to describe models in which non-stationary behaviour is reflected in changing covariances. However, it is worth noting that a trend in the variance can often be removed by a simple transformation such as taking logarithms. More generally, the *Box—Cox transform* may be applied. This is defined by

$$y_t^{(\lambda)} = \begin{cases} (y_t^\lambda - 1)/\lambda, & 0 < \lambda \leqslant 1 \\ \log y_t & , & \lambda = 0. \end{cases} \tag{5.3}$$

Despite the generality of the Box—Cox transform, taking logarithms will be sufficient for most purposes. This is particularly so for economic time series, since the first difference of a logarithm yields a close approximation to the growth rate.

Autoregressive-Integrated-Moving Average Processes

One of the simplest examples of an integrated process is the *random walk*

$$y_t = y_{t-1} + \epsilon_t. \tag{5.4}$$

This model has essentially the same form as the first-order autoregressive process, (2.2.2), but the condition that ϕ lies within the unit circle is clearly violated. The result is that (5.4) is not a stationary process. Repeatedly substituting for past values of y_t gives

$$y_t = \sum_{j=0}^{t-1} \epsilon_{t-j} + y_0, \qquad t = 1, 2, \ldots, T. \tag{5.5}$$

The mean of y_t is therefore a function of the initial value, y_0, as $E(y_t) = E(y_0)$. If y_0 is fixed, the first requirement for stationarity, namely that the mean be constant over time, is satisfied. Nevertheless, the process is non-stationary since

$$\text{Var}(y_t) = t\sigma^2 \tag{5.6}$$

and

$$\text{Cov}(y_t, y_{t-\tau}) = |t - \tau| \sigma^2. \tag{5.7}$$

The random walk process therefore tends to meander away from its starting value, but exhibits no particular trend in doing so. Despite this behaviour a stationary process is very easily obtained. Re-arranging (5.4) gives

$$\Delta y_t = y_t - y_{t-1} = \epsilon_t \tag{5.8}$$

and so the first differences follow a white noise process.

An obvious extension of (5.8) is to replace ϵ_t by a general stationary ARMA(p, q) process. On the other side, y_t may be differenced any number of times. If d denotes the degree of differencing the resulting model is

$$\phi(L)\Delta^d y_t = \theta(L)\epsilon_t. \tag{5.9}$$

This is known as an ARIMA(p, d, q) process. For the random walk, $y_t \sim$ ARIMA(0, 1, 0). Stationary ARMA processes are a special case in which $d = 0$.

Model (5.9) may be extended by adding a constant term with parameter β. This gives

$$\Delta^d y_t = \beta + \phi^{-1}(L)\theta(L)\epsilon_t. \tag{5.10}$$

Adding a constant term to a stationary ARMA process permits it to have a non-zero mean without materially altering its behaviour. The effect on an integrated process is more dramatic. Adding a constant term to (5.4) yields

$$\Delta y_t = \beta + \epsilon_t. \tag{5.11}$$

This is sometimes known as a *random walk with drift*. Repeatedly substituting for lagged values of y_t and taking expectations gives

$$E(y_t) = y_0 + \beta t \tag{5.12}$$

if y_0 is fixed. Thus the level of y_t is governed by a linear time trend. Such a trend could also be constructed within the framework of (5.1) and (5.2), but the distinguishing feature of (5.11) is that it cannot be decomposed into a linear trend plus a *stationary* disturbance term.

Prediction

A basic feature of the predictions from stationary models is that they tend towards the mean of the series as the lead time, l, increases. If l is large, the structure of the model is irrelevant. This is no longer the case with integrated processes, where the forecast function contains a deterministic component which depends both on the degree of differencing and on the ARMA model fitted to the differenced observations.

The mechanics of making predictions from ARIMA models are exactly the same as those involved in making predictions from ARMA models. The term $\phi(L)\Delta^d$ in (5.9) is expanded to yield an AR polynomial of order $p + d$, i.e.

$$\phi(L)\Delta^d = \varphi(L) = 1 - \varphi_1 L - \cdots - \varphi_{p+d} L^{p+d}, \tag{5.13}$$

Predictions are made from the difference equation

$$\tilde{y}_{T+l/T} = \varphi_1 \tilde{y}_{T+l-1/T} + \cdots + \varphi_{p+d} \tilde{y}_{T+l-p-d/T}$$
$$+ \tilde{\epsilon}_{T+l/T} + \cdots + \theta_2 \tilde{\epsilon}_{T+l-q/T}, \qquad l = 1, 2, \ldots \tag{5.14}$$

where $\tilde{y}_{T+j/T}$ and $\tilde{\epsilon}_{T+j}$ are defined as in (4.2).

Example 1 In the ARIMA$(0, 1, 1)$ model,

$$\Delta y_t = \epsilon_t + \theta \epsilon_{t-1}, \tag{5.15}$$

forecasts are constructed from the difference equation

$$\tilde{y}_{t+l/T} = \tilde{y}_{T+l-1/T} + \tilde{\epsilon}_{T+l/T} + \theta \tilde{\epsilon}_{T+l-1/T}, \qquad l = 1, 2, \ldots \tag{5.16}$$

Since the expectation of future values of ϵ_t is zero, it follows that

$$\tilde{y}_{T+1/T} = y_T + \theta \epsilon_T \tag{5.17a}$$

and

$$\tilde{y}_{T+l/T} = \tilde{y}_{T+l-1/T}, \qquad l = 2, 3, \ldots \tag{5.17b}$$

Thus, for all lead times, the forecasts made at time T follow a horizontal straight line.

The disturbance term, ϵ_T, is constructed from the difference equation

$$\epsilon_t = y_t - y_{t-1} - \theta \epsilon_{t-1}, \qquad t = 2, \ldots, T \tag{5.18}$$

with $\epsilon_1 = 0$, cf. the CSS recursion given in (5.3.3). Substituting repeatedly for lagged values of ϵ_t in the equation for ϵ_T gives

$$\epsilon_T = y_T + (1 + \theta) \sum_{j=1}^{T-1} (-\theta)^j y_{T-j}. \tag{5.19}$$

The expression for the one-step ahead predictor, (5.17a), can therefore be written as

$$\tilde{y}_{T+1/T} = (1 + \theta) \sum_{j=0}^{T-1} (-\theta)^j y_{T-j}. \tag{5.20}$$

Hence $\tilde{y}_{T+1/T}$ is an *exponentially weighted moving average* (EWMA) of current and past observations. Such a forecast has an obvious intuitive appeal. Furthermore, it is very easy to implement. Expression (5.20) can be re-arranged to give

$$\tilde{y}_{T+1/T} = \lambda y_T + (1 - \lambda)\tilde{y}_{T/T-1}, \tag{5.21}$$

where $\lambda = 1 + \theta$, thereby showing that the current forecast is a weighted average of the current observation and the previous forecast.

The simplicity of the EWMA forecasting procedure has made it an attractive proposition in situations where forecasts must be made for a large number of series and the resources for building a sophisticated model are limited. Such situations arise frequently in operational research in connection with stocks. The *smoothing constant*, λ, is assigned a value between zero and one. If λ is close to zero, the forecast relies heavily on past observations. If $\lambda = 1$ the evidence of past data is ignored completely and the forecast is given by the value of the current observation.

Example 2   The eventual forecast function in the ARIMA(0, 2, 2) model,

$$\Delta^2 y_t = \epsilon_t + \theta_1 \epsilon_{t-1} + \theta_2 \epsilon_{t-2}, \tag{5.22}$$

is the solution of $(1 - L)^2 \tilde{y}_{T+l/T} = 0$. This is a straight line passing through the first two predictions, $\tilde{y}_{T+1/T}$ and $\tilde{y}_{T+2/T}$. As noted in Section 1.1, the forecasts can be updated very easily, the relevant formulae, (1.1.8), being a generalisation of (5.21).

Example 3   In the ARIMA(1, 1, 0) model,

$$\varphi(L) = (1 - \phi L)(1 - L) = 1 - (1 + \phi)L + \phi L^2$$

and forecasts are constructed from the difference equation

$$\tilde{y}_{T+l/T} = (1 + \phi)\tilde{y}_{T+l-1/T} - \phi\tilde{y}_{T+l-2/T}, \qquad l = 1, 2, \ldots \tag{5.23}$$

The corresponding forecast function is

$$\tilde{y}_{T+l/T} = y_T + (y_T - y_{T-1})\frac{\phi(1 - \phi^l)}{1 - \phi}, \qquad l = 1, 2, \ldots \tag{5.24}$$

As $l \to \infty$ this approaches the horizontal line

$$y_T + (y_T - y_{T-1})\phi/(1 - \phi). \tag{5.25}$$

In the ARIMA$(0, 1, 1)$ model the forecast function is horizontal, but its level depends on all past values; see (5.20). By way of contrast (5.24) depends only on the last two points in the series. The trend which is extrapolated is therefore a very local one. This is a feature of all ARIMA models in which there is no MA component.

The general point to emerge from these examples is that the eventual forecast function for an ARIMA(p, d, q) model is a polynomial of order $d - 1$. If the model has a non-zero mean, the eventual forecast function is a polynomial of order d.

Example 4 In (5.11) the forecast function is

$$\tilde{y}_{T+l/T} = \alpha + \beta l, \qquad l = 1, 2, \ldots \tag{5.26}$$

where $\alpha = y_T$. As was pointed out in Example 2, the eventual forecast function for an ARIMA$(0, 2, 2)$ process *without* a constant term is also a linear trend. The distinguishing feature of (5.26) is that the slope, β, is interpreted as the average incremental increase in y_t over the whole length of the series. In the ARIMA$(0, 2, 2)$ model, the later observations are relatively more important in determining the slope of the forecast function.

Adding a constant term to an integrated process introduces a deterministic component into the series. Box and Jenkins (1976, pp. 92–3) argue that this is unrealistic in most situations. They prefer to assume that $\beta = 0$ 'unless such an assumption proves contrary to the facts presented by the data.'

The MSE of a prediction from an ARIMA model may be obtained from formula (4.12), the ψ coefficients being calculated by equating powers of L in the expression

$$\varphi(L)\psi(L) = \theta(L). \tag{5.27}$$

This yields

$$\psi_j = \sum_{i=1}^{\min(j, p+d)} \varphi_i \psi_{j-i} + \theta_j, \qquad 1 \leqslant j \leqslant q \tag{5.28a}$$

$$\psi_j = \sum_{i=1}^{\min(j, p+d)} \varphi_i \psi_{j-i}, \qquad j > q. \tag{5.28b}$$

Note that $\psi_0 = 1$.

Example 5 In the ARIMA$(0, 2, 1)$ model

$$\Delta^2 y_t = \epsilon_t + 0.5\epsilon_{t-1} \tag{5.29}$$

the ψ_j's are calculated as follows:

$$\psi_1 = \varphi_1 + \theta_1 = 2.0 + 0.5 = 2.5$$

$$\psi_2 = \varphi_1 \psi_1 + \phi_2 = 2(2.5) - 1.0 = 4.0$$

$$\psi_3 = \varphi_1 \psi_2 + \varphi_2 \psi_1 = 8.0 - 2.5 = 5.5.$$

Thus for $l = 3$

$$\text{MSE}(\tilde{y}_{T+3/T}) = \sigma^2 (1.0^2 + 2.5^2 + 4.0^2) = 23.25\,\sigma^2$$

and the 95% prediction interval for y_{T+3} is

$$\tilde{y}_{T+3/T} \pm 9.5\sigma.$$

The MSE's of forecasts in ARIMA models tend to increase rapidly as the lead time becomes greater. This stresses the point that the main value of such models is in short-term forecasting.

Stochastic Trend Plus Error Models

Further insight into the structure of ARIMA models can be obtained by replacing the deterministic trend in (5.1) by a stochastic trend, m_t. In order to qualify as a trend, m_t must be a non-stationary process.

Example 6   Suppose that m_t is a random walk

$$m_t = m_{t-1} + \eta_t \tag{5.30}$$

with $\eta_t \sim \text{WN}(0, \sigma_\eta^2)$, and that the 'error' term is a white noise process, ϵ_t, with variance σ^2. The model is then

$$y_t = m_t + \epsilon_t, \qquad t = 1, \ldots, T \tag{5.31}$$

and taking first differences yields

$$\Delta y_t = \eta_t + \epsilon_t - \epsilon_{t-1}, \qquad t = 2, \ldots, T. \tag{5.32}$$

This is an ARIMA(0, 1, 1) process of the form

$$\Delta y_t = \xi_t + \theta \xi_{t-1}, \qquad t + 2, \ldots, T, \tag{5.33}$$

where ξ_t is white noise; cf. Example 2.5.2. Equating the auto-correlations at lag one in the two models yields $\kappa = -(1 + \theta)^2/\theta$, where κ is the variance ratio σ_η^2/σ^2. This is a quadratic equation, the root corresponding to an invertible model being

$$\theta = [(\kappa^2 + 4\kappa)^{1/2} - 2 - \kappa]/2. \tag{5.34}$$

It was shown in Example 1 that the optimal forecast for an ARIMA(0, 1, 1) model takes the form of an exponentially weighted moving average of past observations. The model in this example provides a rationale for the EWMA as the appropriate forecasting procedure for a series in which the mean is slowly changing over time.

Example 7 The stochastic trend in (5.31) can be generalised by letting m_t be a random walk with drift, in which the drift component is itself random. Thus

$$m_t = m_{t-1} + \beta_t + \eta_t \qquad\qquad (5.35\text{a})$$

$$\beta_t = \beta_{t-1} + \zeta_t, \qquad\qquad (5.35\text{b})$$

where $\zeta_t \sim \text{WN}(0, \sigma_\zeta^2)$. Taking second differences gives

$$\Delta^2 y_t = (\eta_t + \zeta_t + \epsilon_t) - (\eta_{t-1} + 2\epsilon_{t-1}) + \epsilon_{t-2}. \qquad (5.36)$$

This is an ARIMA(0, 2, 2) process, and forecasts can be constructed from the recursive formulae, (1.1.8), the smoothing constants depending on the relative variances of the disturbance terms. The model defined by (5.31) and (5.35) therefore provides a rationale for the Holt—Winters forecasting procedure.

6. Seasonality

When observations are available on a monthly or a quarterly basis, some allowance must be made for seasonal effects. One approach would be to work with seasonally adjusted data, but for the reasons discussed in the final sub-section this is not generally desirable. The bulk of this section is therefore concerned with ways in which seasonality can be incorporated into time series models.

There are two aspects to seasonality. The first is concerned with patterns in the observations which are repeated more or less regularly from year to year. These features are permanent, and they may be regarded as stemming from factors such as the weather. Modelling such effects involves considerations similar to those which arise in modelling trends. A deterministic seasonal component is appropriate if it is felt that the seasonal pattern is constant from year to year, while a stochastic component can be introduced if it is felt that the pattern is slowly changing. Both models project a regular seasonal pattern into the future, but when a stochastic model is used this pattern depends more heavily on the later observations in the series. There is a direct parallel with the concept of a local trend in a non-seasonal series.

The other aspect of seasonality concerns effects which arise in the absence of, or in addition to, the type of seasonality described in the previous paragraph. For example, a dock strike in March of last year could influence production targets in March of this year, if firms believe that there is a high probability of such an event happening again. However, unless dock strikes in March turn out to be a regular occurrence, the effect of the original strike will be transitory. Thus, while it is likely that observations in the same seasons of different years will be correlated, these correlations can be expected to be small if the years are a long way apart. These considerations suggest a stationary model, in which any seasonal pattern tends to disappear as the lead time of a forecast increases. Models of this kind are discussed in the first sub-section below. This is followed by a description of models with deterministic and slowly changing seasonal effects.

Seasonal Autoregressive-Moving Average Processes

Consider a stationary series of quarterly observations. A very simple way of capturing a seasonal effect is by a fourth-order seasonal autoregressive process of the form

$$y_t = \phi_4 y_{t-4} + \epsilon_t, \qquad |\phi_4| < 1. \tag{6.1}$$

This model is a special case of an AR(4) process, but with the constraint that $\phi_1 = \phi_2 = \phi_3 = 0$. The properties of (6.1) are very similar to those of a stationary AR(1) process and using the techniques developed in Section 2.2, it is not difficult to show that the autocorrelation function is given by

$$\rho(\tau) = \begin{cases} \phi_4^{\tau/4}, & \tau = 0, 4, 8, \ldots \\ 0, & \text{otherwise.} \end{cases} \tag{6.2}$$

The closer $|\phi_4|$ is to unity, the stronger the seasonal pattern. However, as long as ϕ_4 remains within the unit circle, the seasonal effect is non-deterministic, with the seasonal pattern gradually dying away as predictions are made further and further ahead.

Model (6.1) may be extended to allow for both AR and MA terms at other seasonal lags. If s denotes the number of seasons in the year, a general formulation is as follows:

$$\phi^\dagger(L^s)y_t = \theta^\dagger(L^s)\zeta_t, \tag{6.3}$$

where

$$\phi^\dagger(L^s) = 1 - \phi_1^\dagger L^s - \cdots - \phi_P^\dagger L^{Ps} \tag{6.4a}$$

$$\theta^\dagger(L^s) = 1 + \theta_1^\dagger L^s + \cdots + \theta_Q^\dagger L^{Qs} \tag{6.4b}$$

and ζ_t is a white noise disturbance term. The value of s will typically be either four or twelve in economic applications, corresponding to quarterly and monthly observations respectively.

Expression (6.3) defines a pure seasonal ARMA process of order $(P, Q)_s$. As in the case of (6.1), the autocorrelation function will contain 'gaps' at non-seasonal lags. However, unless seasonal movements are felt to be the only predictable feature of the series, such a model will not be appropriate. With monthly data, for example, it seems reasonable to suppose that an observation in March will be related to the observation in February, as well as to the observation in March of the previous year.

There are basically two ways of constructing models which make allowance for both seasonal and non-seasonal movements. The first is simply to fill in the gaps in the seasonal process. Thus a first-order lag might be incorporated into (6.1) to yield

$$y_t = \phi_1 y_{t-1} + \phi_4 y_{t-4} + \epsilon_t. \tag{6.5}$$

A second way is to replace the white noise disturbance term in (6.3) by a non-seasonal ARMA(p, q) process, u_t; i.e.

$$\phi(L)u_t = \theta(L)\epsilon_t, \tag{6.6}$$

where $\phi(L)$ and $\theta(L)$ are the associated polynomials defined in (2.1.26). Combining (6.3) and (6.6) produces

$$\phi^\dagger(L^s)\phi(L)y_t = \theta^\dagger(L^s)\theta(L)\epsilon_t. \tag{6.7}$$

This is a *multiplicative* seasonal ARMA process of order $(p, q) \times (P, Q)_s$. As a simple example, suppose that an AR(1) process is incorporated into (6.1). The model becomes

$$(1 - \phi_1 L)(1 - \phi_4 L^4) = \epsilon_t. \tag{6.8}$$

This can be re-written as

$$y_t = \phi_1 y_{t-1} + \phi_4 y_{t-4} + \phi_5 y_{t-5} + \epsilon_t \tag{6.9}$$

where $\phi_5 = -\phi_1 \phi_4$.

Multiplicative forms are widely used in time series modelling. However, there is nothing in the development so far to indicate why, for example, (6.9) should be preferred to (6.5). It will be argued later that multiplicative models arise most naturally when differencing has taken place.

Although the methods described above are the two most common ways of formulating stationary seasonal models, there is a third possibility. This is to construct an *additive* model

$$y_t = s_t + u_t, \tag{6.10}$$

in which u_t is an ARMA(p, q) process of the form (6.6) and s_t is a seasonal ARMA process of order $(P, Q)_s$. The disturbance terms driving the two processes are assumed to be independent. As an example consider

$$y_t = \frac{\zeta_t}{1 - \phi_4 L} + \frac{\epsilon_t}{1 - \phi_1 L}. \tag{6.11}$$

This can be re-written as

$$(1 - \phi_1 L)(1 - \phi_4 L^4) y_t = \zeta_t + \epsilon_t - \phi_1 \zeta_{t-1} - \phi_4 \epsilon_{t-4}. \tag{6.12}$$

The right-hand side of (6.12) is an MA(4) process, but the parameters are subject to restrictions determined by ϕ_1 and ϕ_4. One way of taking account of these restrictions in estimating the model is to write (6.11) in state space form and compute the likelihood function using the Kalman filter, cf. Example 4.4.3. An alternative approach is described in Nerlove *et al.* (1979).

Deterministic Seasonal Models

Let u_t be a zero mean stationary process which may or may not include seasonal effects of the type described in the previous sub-section. A deterministic seasonal pattern may be superimposed on this series by introducing a set of s dummy variables into the model. If z_{jt} is a dummy variable taking a value of unity in season j and a value of zero otherwise, the model is

$$y_t = \sum_{j=1}^{s} \gamma_j z_{jt} + u_t, \qquad t = 1, \ldots, T. \tag{6.13}$$

The jth seasonal parameter, γ_j, can be interpreted as the expectation of y_t in the jth season.

An alternative way of modelling a deterministic seasonal pattern is to use a set of trigonometric terms. Equation (6.13) is replaced by

$$y_t = \alpha_0 + \sum_j (\alpha_j \cos \lambda_j t + \beta_j \sin \lambda_j t) + u_t, \tag{6.14}$$

where the summation is from $j = 1$ to $s/2$ if s is even, and from $j = 1$ to $(s - 1)/2$ if s is odd, and $\lambda_j = 2\pi j/s$.

Equation (6.14) contains s seasonal parameters and can therefore represent a seasonal pattern in exactly the same way as the s dummy variables in (6.13). The advantage of the trigonometric representation is that it will often be possible to describe the seasonal pattern using less than s coefficients. This consideration may be particularly important for monthly data. In the butter market example quoted in

Section 3.2, the two lowest frequency components, corresponding to periods of twelve and six months, accounted for over 90% of the seasonal variation. It would probably be reasonable to include only these two frequencies in the seasonal model, thereby reducing the number of seasonal parameters from twelve to five.

Changing Seasonal Patterns

The model in (6.13) may be written in the form

$$y_t = s_t + u_t, \qquad t = 1, \ldots, T. \tag{6.15}$$

In the jth season of any year the deterministic component, s_t, is simply equal to γ_j and so

$$s_t = s_{t-s} \tag{6.16}$$

for all t. A degree of flexibility may be introduced into the seasonal pattern by adding to (6.16) a disturbance term, ω_t, which is independent of u_t. This yields

$$s_t = s_{t-s} + \omega_t. \tag{6.17}$$

It is equivalent to letting the γ_j's be random variables.

Letting s_t follow a random walk of the kind defined in (6.17) allows the seasonal pattern to change slowly over time. Furthermore, although s_t is non-stationary, a stationary process can be obtained by *seasonal differencing*. Thus

$$\Delta_s y_t = \omega_t + u_t - u_{t-s} \tag{6.18}$$

where Δ_s is the seasonal difference operator

$$\Delta_s = 1 - L^s. \tag{6.19}$$

Provided the variance of ω_t is strictly positive, the right-hand side of (6.18) is invertible.

Example 1   If u_t and ω_t are white noise processes, it follows from an argument similar to the one employed in Example 5.6 that $\Delta_s y_t$ is a seasonal MA(1) process. For monthly data,

$$\Delta_{12} y_t = \xi_t + \theta_{12} \xi_{t-12}, \tag{6.20}$$

where ξ_t is white noise.

Although the seasonal component modelled by (6.17) is stochastic, a fixed seasonal pattern is projected into the future. For $l \geqslant 13$ the forecast function for (6.20) is

$$\tilde{y}_{T+l/T} = \tilde{y}_{T+l-12/T}, \tag{6.21}$$

the starting values $\bar{y}_{T+1/T}$ to $\bar{y}_{T+12/T}$ depending more heavily on the later observations than the earlier one. In the extreme case where $u_t \equiv 0$ for all t, $\theta_{12} = 0$ and it is the pattern of the last twelve observations which is projected into the future.

A somewhat smoother seasonal pattern will be obtained if ω_t exhibits positive serial correlation. This can be achieved by it following a non-seasonal ARMA process.

Example 2   Permanent and transitory seasonal components can be combined. Suppose that the observations are quarterly, and that y_t is the sum of a permanent component, (6.17), and a stationary seasonal AR(4) process of the form (6.1). Then

$$(1 - \phi_4 L)(1 - L^4)y_t = \omega_t - \phi_4 \omega_{t-4} + \epsilon_t - \epsilon_{t-4} \qquad (6.22)$$

and so $\Delta_4 y_t$ follows a seasonal ARMA process of order $(1, 1)_4$.

Seasonal Adjustment

Official statistics are frequently published as seasonally adjusted series. There is a tendency to believe that working with seasonally adjusted data simplifies matters, in that it becomes unnecessary to make any explicit allowance for seasonal effects. However, it was shown in Section 3.5 that the procedures adopted by official statisticians tend to result in over-adjustment, insofar as too much power is removed from the spectrum at seasonal frequencies. This phenomenon is also reflected in the time domain, where there is a tendency for a seasonally adjusted series to exhibit negative auto-correlations at the seasonal lags. In the early seventies the official adjustment procedure in the UK was a modification of the US Bureau of the Census Method II, Variant X-11 method. Wallis (1974) examined the effect of this procedure for quarterly data using a number of series, including white noise. The theoretical auto-correlation function for a seasonally adjusted white noise process is shown in figure 6.2. Ideally, the adjusted series should display the characteristics of a white noise process, but the small positive autocorrelations at lags of 1 to 3, 5 to 7, . . . , and the somewhat more pronounced negative autocorrelations at lags of 4, 8, . . . , suggests that this may not be the case. Indeed, it is quite conceivable that one could end up fitting an AR(4) model to the seasonally adjusted data, this being a reasonable approximation to what is essentially a complicated moving average.

Wallis also considered the effect of seasonal adjustment on a seasonal AR process of the form (6.1). In this case he noted a reduction in the autocorrelations at the seasonal lags, but this was

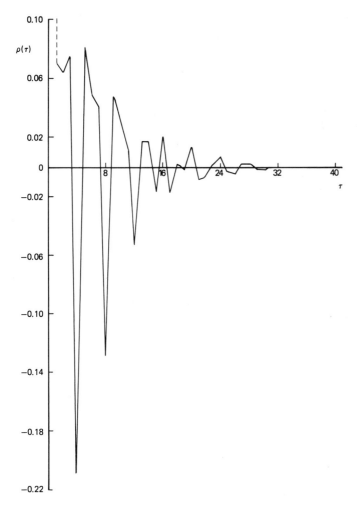

Figure 6.2   *Autocorrelation Function of a Seasonally Adjusted White Noise*
Process

Source: Wallis (1974).

accompanied by the introduction of 'substantial autocorrelation' at
all other lags, cf. (6.2). In certain circumstances he concluded that
this could lead to an AR(1) scheme being adopted for the adjusted
series.

The conclusion which emerges from the work of Wallis, and of
others, is that it is generally preferable to work with the unadjusted
series. Seasonal adjustment can introduce considerable distortion

into a series, and, at the same time, there is no guarantee that the adjusted series will be free from seasonal effects.

7. A General Class of Models

The material presented in the last two sections suggests a general approach to model building based on differencing. This is the philosophy adopted by Box and Jenkins (1976). They propose that the conventional and seasonal differencing operators be applied until the series is stationary and that this stationary series be modelled by a multiplicative seasonal ARMA process of the type defined in (6.7). Such a model may be written

$$\phi^\dagger(L^s)\phi(L)\Delta^d\Delta_s^D y_t = \theta^\dagger(L^s)\theta(L)\epsilon_t \qquad (7.1)$$

where D and d are integers denoting the number of times the seasonal and first difference operators are applied respectively. It is known as a *multiplicative seasonal ARIMA* process of order $(p, d, q) \times (P, D, Q)_s$.

A more general formulation is obtained by relaxing the assumption that the stationary part of the model is a multiplicative process. A model of this kind may be written

$$\phi(L)\Delta^d\Delta_s^D y_t = \theta(L)\epsilon_t, \qquad (7.2)$$

where the polynomials $\phi(L)$ and $\theta(L)$ will typically contain parameters at both seasonal and non-seasonal lags, but may have gaps; cf. (6.5). The multiplicative model is a special case of (7.2) in that the parameters in (7.2) would be subject to constraints if (7.1) were to hold.

Neither of the above models presents any new problems as regards estimation or prediction. The main difficulty lies in model selection. It should, however, be noted that the evidence in Ansley and Newbold (1980) suggests that the case for using exact ML is even stronger than it is for non-seasonal models.

Example 1   The airline passenger series given in Box and Jenkins (1976, p. 531) has become something of a classic in time series modelling. The data relate to monthly totals of passengers over the period January 1949 to December 1960. Figure 6.3 shows a plot of the logarithms of these figures (in thousands), together with the forecasts made from July 1957 using the model

$$\Delta\Delta_{12}y_t = (1 + \theta_1 L)(1 + \theta_{12}L^{12})\epsilon_t. \qquad (7.3)$$

This is a multiplicative seasonal ARIMA process of order $(0, 1, 1) \times (0, 1, 1)_{12}$, with y_t in logarithms. The estimates

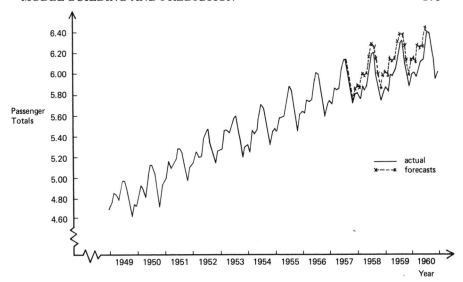

Figure 6.3 *Logarithms of Monthly Airline Passenger Totals (in Thousands)*

Source of Data: Box and Jenkins (1976, p. 531).

obtained by exact ML, using the full set of observations are

$$\tilde{\theta}_1 = -0.40 \qquad \tilde{\theta}_{12} = -0.55 .$$
$$\quad\;\,(0.09) \qquad\qquad (0.07)$$

The figures in parenthesis are asymptotic standard errors. The modified Box–Pierce statistic based on the first 48 residual autocorrelations is $Q^*(46) = 44.12$. This gives no indication of misspecification.

A model of the form (7.2) corresponding to the structure adopted in (7.3) would have MA parameters at lags 1, 12, and 13. However, if (7.2) were the true model, θ_{13} would satisfy the non-linear constraint $\theta_{13} = \theta_1 \theta_{12}$.

Expanding and re-arranging (7.3) with its fitted values yields

$$y_t = y_{t-1} + y_{t-12} - y_{t-13} + \epsilon_t - 0.40\epsilon_{t-1} - 0.55\epsilon_{t-12}$$
$$+ 0.22\epsilon_{t-13} \tag{7.4}$$

and predictions may be built up recursively from this equation in the usual way. The eventual forecast function is a linear trend with a regular seasonal pattern superimposed on top of it.

Trend, Seasonal and Error Components

The rationale for the seasonal ARIMA form can be seen by considering an additive model consisting of trend, seasonal and error components. This may be written

$$y_t = m_t + s_t + u_t, \qquad t = 1, \ldots, T. \tag{7.5}$$

As such it is a generalisation of (5.31) and (6.15).

If m_t is a random walk with drift,

$$m_t = \beta + m_{t-1} + \eta_t, \tag{7.6}$$

and s_t is a stochastic seasonal component of the form, (6.17), a stationary process is obtained by taking first differences followed by seasonal differences. Thus

$$\Delta\Delta_s y_t = (1 - L^s)\eta_t + (1 - L)\omega_t + (1 - L)(1 - L^s)u_t,$$
$$t = s + 1, \ldots, T. \tag{7.7}$$

When u_t is a white noise disturbance term it can be denoted by ϵ_t and (7.7) can be re-written as

$$\Delta\Delta_s y_t = (\eta_t + \omega_t + \epsilon_t) - (\omega_{t-1} + \epsilon_{t-1}) - (\eta_{t-s} + \epsilon_{t-s}) + \epsilon_{t-s-1}. \tag{7.8}$$

The right-hand side of (7.8) is a multiplicative seasonal ARMA process of the form (6.7). The model as a whole is therefore a seasonal ARIMA process of order $(0, 1, 1) \times (0, 1, 1)_s$, i.e.

$$\Delta\Delta_s y_t = (1 + \theta_1 L)(1 + \theta_s L^s)\xi_t, \tag{7.9}$$

where ξ_t is white noise.

The link between (7.8) and (7.9) may be established by examining their respective autocorrelation functions. For the seasonal ARIMA model

$$\rho(1) \quad = \theta_1/(1 + \theta_1^2) \tag{7.10a}$$

$$\rho(s - 1) = \theta_1 \theta_s/(1 + \theta_1^2)(1 + \theta_s^2) \tag{7.10b}$$

$$\rho(s) \quad = \theta_s/(1 + \theta_s^2) \tag{7.10c}$$

$$\rho(s + 1) = \rho(s - 1) \tag{7.10d}$$

The autocorrelations at all other lags are zero. For the model derived from (7.5) the variance is

$$\gamma(0) = 2\sigma_\eta^2 + 2\sigma_\omega^2 + 4\sigma^2. \tag{7.11}$$

The non-zero autocovariances are

$$\gamma(1) \quad = -(\sigma_\omega^2 + 2\sigma^2) \tag{7.12a}$$

$$\gamma(s - 1) = \sigma^2 \tag{7.12b}$$

$$\gamma(s) \quad = -(\sigma_\eta^2 + 2\sigma^2) \tag{7.12c}$$

$$\gamma(s + 1) = \gamma(s - 1). \tag{7.12d}$$

Dividing the autocovariances by $\gamma(0)$ and comparing them with (7.10) yields

$$\bar{\sigma}_\eta^2 = \sigma_\eta^2/\sigma^2 = -(1 + \theta_1)^2/\theta_1 \tag{7.13a}$$

and

$$\bar{\sigma}_\omega^2 = \sigma_\omega^2/\sigma^2 = -(1 + \theta_s)^2/\theta_s. \tag{7.13b}$$

Example 2   The model fitted to the airline data in Example 1 is of the form (7.9). Using (7.13), the estimated relative variances of the trend and seasonal components are 0.90 and 0.37 respectively. The seasonal pattern therefore changes quite slowly.

Overdifferencing

Differencing a series h times will eliminate a deterministic polynomial trend of degree h. However, if the disturbance term in (5.1) is stationary, the disturbance term in the differenced model will be strictly non-invertible, the MA associated polynomial having h roots on the unit circle.

Example 3   Consider the linear trend model

$$y_t = \alpha + \beta t + \epsilon_t. \tag{7.14}$$

Differencing twice gives

$$\Delta^2 y_t = (1 - L)^2 \epsilon_t = \epsilon_t - 2\epsilon_{t-1} + \epsilon_{t-2}. \tag{7.15}$$

Although $\Delta^2 y_t$ is a stationary series with zero mean it is non-invertible. The same result can be demonstrated in terms of the model in Example 5.7. If $\sigma_\eta^2 = \sigma_\zeta^2 = 0$, the stochastic trend becomes a deterministic linear trend and (5.36) is an ARIMA(0, 2, 2) process with $\theta_1 = -2$ and $\theta_2 = 1$.

Seasonal differencing also has a role to play in removing polynomial trends. In addition, of course, seasonal differencing will eliminate a deterministic seasonal component.

Example 4   The model

$$y_t = \alpha + \beta t + \sum_{j=1}^{11} \gamma_j z_{jt} + \epsilon_t, \qquad t = 1, \ldots, T \tag{7.16}$$

is a possible candidate for the airline passenger series. The z_{jt}'s are seasonal dummies, as defined in (6.13), but only eleven are needed as a constant term is already included. Taking first differences followed by twelfth differences yields

$$\Delta\Delta_{12}y_t = (1 - L)(1 - L^{12})\epsilon_t, \qquad t = 13, \ldots, T. \qquad (7.17)$$

This is a multiplicative seasonal ARIMA model of the form, (7.3), with $\theta_1 = \theta_{12} = -1$. Like (7.15), it is non-invertible.

The fact that the ML estimates of θ_1 and θ_{12} in (7.3) are both well inside the unit circle is a reasonable indication that the deterministic model in (7.16) is inappropriate.

If a differenced series is stationary but strictly non-invertible, *overdifferencing* is said to have taken place. A failure to recognise that a series has been overdifferenced is only likely to create problems if an inappropriate estimation procedure is used, or if forecasts are constructed without using a finite sample prediction procedure. The question of estimation was discussed in Section 5. As regards forecasting, nothing is lost if finite sample predictions are made from non-invertible models like (7.15) and (7.17). In fact the predictions for any lead time are identical to the predictions which would have been obtained by estimating (7.14) and (7.16) by ordinary least squares; see Harvey (1981).

Example 5*   Consider a white noise process with mean μ:

$$y_t = \mu + \epsilon_t, \qquad t = 1, \ldots, T. \qquad (7.18)$$

The MMSLE of $y_{T+l/T}$ is $\bar{y}$ and this has an MSE of $(1 + T^{-1})\sigma^2$ for all l.

Now suppose that first differences are taken. If the resulting process is identified as ARIMA(0, 1, 1), it may be put in state space form for $t = 2, \ldots, T$; cf. Example 4.1.1. Applying the Kalman filter yields

$$a_T = \begin{bmatrix} \Delta y_T \\ \theta\tilde{\epsilon}_T \end{bmatrix}, \qquad P_T = \begin{bmatrix} 0 & 0 \\ 0 & f_{T+1} - 1 \end{bmatrix}, \qquad (7.19)$$

where f_t is now defined as

$$f_t = 1 + \theta^{2(t-1)}/[1 + \theta^2 + \cdots + \theta^{2(t-2)}],$$

$$t = 2, \ldots, T+1. \qquad (7.20)$$

Since predictions are needed for y_t, rather than Δy_t, the state space formulation must be modified. One way of doing this is to treat y_t as an ARMA(1, 1) process in which $\phi = 1$. It can be shown that the new state vector at time $t = T$ is given by $a_T^\dagger = (y_T, \theta\tilde{\epsilon}_T)'$ with P_T defined by (7.19). The formulae for multi-step prediction, (4.5.1) and (4.5.2), then yield

$$\tilde{y}_{T+l/T} = y_T + \theta \tilde{e}_T, \qquad l = 1, 2, \ldots \tag{7.21a}$$

and

$$P_{T+1/T} = \begin{bmatrix} f_{T+1} & \theta \\ \theta & \theta^2 \end{bmatrix},$$

$$P_{T+2/T} = \begin{bmatrix} f_{T+1} + 2\theta + \theta^2 + 1 & \theta \\ \theta & \theta^2 \end{bmatrix} \tag{7.21b}$$

and so on. If $\theta = -1$, the top left-hand element of $P_{T+l/T}$ for all lead times is simply f_{T+1}, and so from (7.20) and (4.5.4) the MSE of any l-step ahead forecast is

$$\mathrm{MSE}(\tilde{y}_{T+l/T}) = \sigma^2(1 + T^{-1}), \qquad l = 1, 2, 3, \ldots \tag{7.22}$$

This is identical to the MSE of $\bar{y}$ and, in fact, (7.21a) and $\bar{y}$ turn out to be identical.

Model Selection

Methods of identifying a suitable ARMA model from the correlogram and partial autocorrelation function were described in Section 3. The same techniques can be employed with a seasonal ARIMA model once the degree of differencing has been decided on. The only additional complication concerns the identification of multiplicative terms. One possible approach to this problem is to work initially with non-multiplicative models of the form (7.2) and then to search for constraints. Thus, if a model of the form

$$\Delta\Delta_{12} y_t = \epsilon_t + \theta_1 \epsilon_{t-1} + \theta_{12} \epsilon_{t-12} + \theta_{13} \epsilon_{t-13} \tag{7.23}$$

were identified, the next step would be to determine whether the restriction $\theta_{13} = \theta_1 \theta_{12}$ was plausible. If it was, the multiplicative model, (7.3), could be employed.

There are a number of ways of deciding on the appropriate degree of differencing. The approach favoured by Box and Jenkins is based on an examination of the correlogram. For a stationary process, the autocorrelations tend towards zero as the lag increases, and this feature is reflected in the correlogram. The sample autocorrelations from a non-stationary process, on the other hand, do not typically damp down to zero as the lag increases. A heuristic illustration of why this is the case is provided by considering the stationary AR(1) model. The autocorrelation function, (2.2.30), damps down more slowly the closer $|\phi|$ is to 1. When $\phi = 1$, the process is a random walk and the theoretical autocorrelation function is not defined, but the sample autocorrelation function can still be computed.

If an inspection of the correlogram shows relatively large auto-correlations at high lags, differencing of the observations is probably appropriate. The strategy adopted by Box and Jenkins is to difference until a correlogram having the characteristics of a stationary process is obtained. For monthly or quarterly observations, some experimentation with various combinations of first differences and seasonal differences may be necessary. The use of the correlogram to determine the order of differencing is based on large sample considerations. When T is small the picture may be less clear cut. It was pointed out in Section 3 that for a stationary process, the correlogram may not damp down as quickly as theoretical considerations would suggest because the sample autocorrelations themselves are correlated. On the other hand, the correlograms of non-stationary processes can damp down quite quickly when T is small. When there is some doubt as to the degree of differencing, it may be worth fitting suitable models to the main contenders and making a choice on the basis of the AIC. Incorporating deterministic trend and seasonal components, $f(t)$ and $s(t)$, into (7.2) yields

$$\Delta^d \Delta_s^D y_t = f(t) + s(t) + \phi^{-1}(L)\theta(L)\epsilon_t. \tag{7.24}$$

If the right-hand side of (7.24) contains n parameters, the goodness of fit criterion is

$$\text{AIC}^\dagger = \tilde{\sigma}^2 \exp\{[\log|V| + 2(n + d + sD)]/T\}, \tag{7.25}$$

where $\sigma^2 V$ is the covariance matrix of $\phi^{-1}(L)\theta(L)\epsilon_t$, and T denotes the number of observations *before* any differencing takes place. It is this value of T which is used as the divisor in the formula for $\tilde{\sigma}^2$.

Example 6   Fitting the deterministic model, (7.16), to the airline passenger series gave $\tilde{\sigma}^2 = 0.00328$. The model selection criterion is therefore

$$\text{AIC}^\dagger = 0.00328 \exp(2 \times 13/144) = 384 \times 10^{-5}.$$

For the multiplicative ARIMA model, (7.3), with $\log|V|$ taken to be zero:

$$\text{AIC}^\dagger = 0.00135 \exp(2 \times 15/144) = 166 \times 10^{-5}.$$

The ARIMA model is clearly preferred. Note that (7.3) effectively contains two extra parameters compared with (7.16). This can be rationalised by observing that (7.17) is a special case of (7.3) which yields identical predictions to (7.16) when a finite sample prediction procedure is used.

The superiority of the ARIMA model is confirmed by the

diagnostic checks. The deterministic model would never, in fact, have been entertained as a serious possibility since $Q(8) = 317.6$.

For a non-seasonal model, the order of differencing will usually only be one or two in practice. Setting $d = 3$ will produce a quadratic forecast function and there are few series for which this seems reasonable. However, there are cases where stationarity cannot be achieved by differencing a small number of times. One such case is the trend plus error model, (5.1), in which the trend is a logistic function,

$$f(t) = \alpha/(1 + \beta e^{-\gamma t}), \tag{7.26}$$

where α, β and γ are parameters. This function tends towards an upper limit or saturation level, as $t \to \infty$ and so is extremely useful in certain circumstances. A model based on, say, second differences will yield poor predictions, except perhaps for $l = 1$ or 2.

When the observations are monthly or quarterly, the appropriate seasonal differencing operator will normally be applied. The case for setting D equal to a value greater than one is very weak, and $D = 1$ is usually combined with $d = 0$ or 1. When $d = D = 1$, the eventual forecast function is a linear trend with a fixed seasonal pattern superimposed on top of it.

For diagnostic checking, Box—Pierce and Lagrange multiplier tests can be constructed in the same way as before. A reasonably large number of lags must be employed in the Box—Pierce test in order to include any remaining seasonal effects. Thus Box and Jenkins take $P = 48$ in forming a Q-statistic for the airline data. Unfortunately, the effect of one or two abnormally large autocorrelations can easily be submerged when such a high value of P is used, and so it is particularly important to examine the correlogram visually. It is conceivable that the cumulative periodogram may be a more effective diagnostic check in these circumstances even though the associated test procedure is not valid except when $p = q = 0$.

8. Multivariate Modelling

While many of the techniques applicable to univariate model building can be extended straightforwardly to the multivariate case, a number of additional problems do arise. In particular, identification becomes much more complex. Rather than attempt to give a definitive summary of multivariate model building, this section will focus on a particular application, and the various aspects of the problem will be discussed as they arise.

The study chosen is by Chan and Wallis (1978). The data relate to the number of skins of mink and muskrat traded annually by the Hudson Bay Company in Canada from 1848 to 1909. The reason a multivariate analysis is called for is that there is known to be a prey–predator relationship between the two species, and this directly affects the population dynamics of both of them.

Stationarity

A preliminary analysis of the data suggests that the muskrat observations are stationary in first differences, while the mink are stationary in levels. Chan and Wallis argue that differencing each of a pair of series a different number of times distorts the phase relationship between them. Rather than differencing the mink series as well, they carry out a quadratic regression on both series, and analyse the relationship between the residuals. Whether this is preferable to 'overdifferencing' the mink series is open to debate, although it is clear from previous discussions that if overdifferencing *is* carried out, it should be coupled with exact ML estimation and finite sample prediction.

Identification, Estimation and Diagnostic Checking

Letting y_{1t} and y_{2t} denote the detrended muskrat and mink series respectively, Chan and Wallis fit the following univariate models:

(i) $\quad (1 - 1.03L + 0.68L^2 - 0.39L^3 + 0.34L^4)y_{1t} = \eta_{1t},$ $\qquad$ (8.1)
$\qquad$ (0.12) $\quad$ (0.17) $\quad$ (0.17) $\quad$ (0.13)

$\qquad \tilde{\sigma}_1^2 = 0.0789, \qquad Q(16) = 19.12$

(ii) $\quad (1 - 1.36L + 0.67L^2)y_{2t} = (1 - 0.70L)\eta_{2t},$ $\qquad$ (8.2)
$\qquad$ (0.13) $\quad$ (0.09) $\qquad\qquad$ (0.15)

$\qquad \tilde{\sigma}_2^2 = 0.0605, \qquad Q(17) = 15.86.$

These models suggest a number of possible specifications for a multivariate process. Although the model chosen for y_{1t} was a pure AR process, the vector ARMA(2, 1) was found to be the most satisfactory formulation on the basis of LR tests. The fitted model is:

$$\begin{bmatrix} 1 - 1.22L + 0.61L^2 & 0 \\ (0.16) \quad (0.12) & \\ & \\ 0 & 1 - 1.29L + 0.62L^2 \\ & (0.15) \quad (0.13) \end{bmatrix} \begin{bmatrix} y_{1t} \\ \\ \\ y_{2t} \end{bmatrix}$$

$$= \begin{bmatrix} 1 - 0.15L & -0.83L \\ (0.22) & (0.16) \\ \\ 0.37L & 1 - 0.81L \\ (0.14) & (0.14) \end{bmatrix} \begin{bmatrix} \epsilon_{1t} \\ \\ \epsilon_{2t} \end{bmatrix}, \qquad (8.3)$$

with

$$\tilde{\Omega} = \begin{bmatrix} 0.061 & 0.022 \\ 0.022 & 0.053 \end{bmatrix}, \qquad Q(30) = \begin{bmatrix} 28.05 & 18.15 \\ 15.73 & 22.45 \end{bmatrix}.$$

Joint estimation reduces the residual variances below those obtained in the separate univariate models. Furthermore, $|\tilde{\Omega}| = 0.00275$, while for the two univariate models $|\tilde{\Omega}_0| = \sigma_1^2 \times \sigma_2^2 = 0.0789 \times 0.0605 = 0.00477$. For the multivariate case, the AIC suggests a comparison based on

$$\text{AIC}^\dagger = |\Omega| \exp(2n/T). \qquad (8.4)$$

The multivariate model is clearly superior on this criterion since $n = 11$ and $\text{AIC}^\dagger = 0.00397$, while for the univariate models $n = 9$ and $\text{AIC}^\dagger = 0.00643$. (Chan and Wallis do not consider the AIC in their analysis.)

The Q matrix given in (8.3) is a generalisation of the Box–Pierce statistic, the off-diagonal elements being constructed from residual cross-correlations. This matrix may be used as an overall check on the goodness of fit of the model. The distribution of Q has not been formally derived, but if each set of correlations is computed up to a lag of P, a rough indication of goodness of fit is obtained by testing each individual statistic against a χ^2_{P-p-q} distribution. A formal test procedure, based on a 'multivariate portmanteau statistic', has been proposed recently by Hosking (1980).

The two AR polynomials in 8.3 are very similar, and estimating the model subject to the restriction that they are identical, as implied by (2.7.26), yields:

$$(1 - 1.28L + 0.63L^2)y_t = \begin{bmatrix} 1 - 0.27L & -0.79L \\ (0.16) & (0.14) \\ \\ 0.34L & 1 - 0.75L \\ (0.11) & (0.12) \end{bmatrix} \begin{bmatrix} \epsilon_{1t} \\ \\ \epsilon_{2t} \end{bmatrix},$$
$$(0.13) \quad (0.11)$$
$$(8.5)$$

with

$$\tilde{\Omega} = \begin{bmatrix} 0.061 & 0.023 \\ 0.023 & 0.054 \end{bmatrix}.$$

The null hypothesis that the AR operators are the same is not rejected by an LR test.

The relationship between (2.7.23) and (2.7.25) suggests that the only vector ARMA process which has as its solution a set of final equations of the form (8.5) is a first-order autoregression. The fitted model is:

$$
\begin{bmatrix}
1 - 0.79L & 0.68L \\
(0.07) & (0.09) \\
-0.29L & 1 - 0.51L \\
(0.07) & (0.09)
\end{bmatrix} y_t = \epsilon_t
\tag{8.6}
$$

with

$$
\tilde{\Omega} = \begin{bmatrix} 0.061 & 0.022 \\ 0.022 & 0.058 \end{bmatrix}, \qquad Q(30) = \begin{bmatrix} 29.11 & 21.02 \\ 18.41 & 26.38 \end{bmatrix}.
$$

Although $|\tilde{\Omega}| = 0.00305$, which is greater than the corresponding figure for (8.5), model (8.6) has fewer parameters. The reason for this is that the polynomials $|\Phi(L)|$ and $\Phi^\dagger(L)$ are both derived from $\Phi(L)$, yet the implicit restrictions imposed on their coefficients are not taken account of in estimating (8.5). Thus, two fewer parameters are contained in (8.6) and the AIC† is lower.

It is a useful exercise to derive the final equations corresponding to (8.6). Since

$$
\Phi(L) = \begin{bmatrix} 1 - 0.79L & 0.68L \\ -0.29L & 1 - 0.51L \end{bmatrix},
\tag{8.7}
$$

the determinantal polynomial is

$$
|\Phi(L)| = (1 - 0.79L)(1 - 0.51L) - (0.68L)(-0.29L)
$$
$$
= 1 - 1.30L + 0.60L^2.
\tag{8.8}
$$

This is fairly close to the *fitted* AR polynomial in (8.5). The adjoint matrix is

$$
\Phi^\dagger(L) = \begin{bmatrix} 1 - 0.51L & -0.68L \\ 0.29L & 1 - 0.79L \end{bmatrix}.
\tag{8.9}
$$

This may be compared directly with the MA matrix in (8.5), since in terms of (2.7.25), $\Theta(L) = I$ in (8.6). The parameters are quite similar.

The Q-statistic in (8.6) appears to be satisfactory, and in addition Chan and Wallis report that the model, despite its simplicity, passes various 'overfitting tests'. They observe that it provides '. . . a direct account of the interactions of the series, and in doing so captures the essential phenomena noted by biologists'. The off-diagonal coefficients imply that an increase in muskrat is followed by an increase in mink a year later, and an increase in mink is followed by a decrease in muskrat a year later.

*Properties of the Models**

The roots of the AR polynomial in (8.5) are a complex conjugate pair, implying damped oscillations with a period of 9.93 years; see formula (3.4.11). Similar oscillations may be expected in (8.6). The best way of examining the stochastic properties of (8.6) is by computing the spectral density matrix. Chan and Wallis report that the spectra of the two series exhibit considerable power at frequencies corresponding to the ten-year cycle, and that the phase diagram indicates that muskrat leads mink by 2–4 years in this frequency range.

Prediction

Prediction for a vector ARMA process may be carried out in exactly the same way as in the univariate case. The recursion (4.2) is generalised directly to the multivariate situation simply by replacing scalars by vectors and matrices.

For the vector AR(1) model in (8.6), the l-step ahead prediction is

$$\tilde{y}_{T+l/T} = \begin{bmatrix} 0.79 & -0.68 \\ 0.29 & 0.51 \end{bmatrix} \tilde{y}_{T+l-1/T}, \qquad l = 1, 2, \ldots,$$

with $\tilde{y}_{T/T} = y_T$. This is a direct generalisation of (4.4). Since the process is stationary, $\Phi^l \to 0$ as $T \to \infty$. The predictions emerge as damped oscillations. Taking $y_{1T} = y_{2T} = 1$,

$$\tilde{y}_{T+1/T} = \begin{bmatrix} 0.79 - 0.68 \\ 0.29 + 0.51 \end{bmatrix} = \begin{bmatrix} 0.11 \\ 0.80 \end{bmatrix},$$

$$\tilde{y}_{T+2/T} = \begin{bmatrix} 0.79 \times 0.11 - 0.68 \times 0.80 \\ 0.29 \times 0.11 + 0.51 \times 0.80 \end{bmatrix} = \begin{bmatrix} -0.46 \\ 0.44 \end{bmatrix}$$

and so on.

Exactly the same predictions would be obtained by working with the final equations obtained by solving (8.6). Since this is a vector ARMA(2, 1) process, the predictions for each series are governed by the same difference equation, based on (8.8). This makes it clear exactly why the period of the oscillations in the predictions of the two series is identical.

The covariance matrix of the prediction errors may also be computed by generalising the results of Section 4. By repeated substitution

$$\tilde{y}_{T+l/T} = \Phi^l y_T + \sum_{j=0}^{l-1} \Phi^j \epsilon_{T+l-j}, \qquad l = 1, 2, \ldots \tag{8.10}$$

Therefore, conditional on Φ, the MSE of $\tilde{y}_{T+l/T}$ is

$$\text{MSE}(\tilde{y}_{T+l/T}) = \sum_{j=0}^{l-1} \Phi^j \Sigma \Phi'^j. \tag{8.11}$$

Estimating Φ introduces a term of $0(T^{-1})$ into this expression; see Baillie (1979).

Notes

Section 3   LM tests for ARMA models are derived in Godfrey (1978) and discussed further in Poskitt and Tremayne (1980; 1981). Identification using the AIC is discussed in Akaike (1974) and Hannan (1980).

Section 7   The relationship between multiplicative ARIMA models and additive components models is explored in Harrison (1967); Cleveland and Tiao (1976); Engle (1978a); and Nerlove *et al.* (1979). Roy (1977), Hasza (1980) and Fuller (1976, p. 367) examine the correlogram for non-stationary processes. Harvey (1980) discusses the use of the AIC to discriminate between regression models in levels and first differences; ARIMA models are a special case.

Exercises

1. Given the information in Exercise 5.7, how would you test the hypothesis that $\rho(2) = 0$?

2. Find expressions for forecasts 1, 2 and 3 time periods ahead for an ARMA$(0, 2, 3)$ model. Show that the forecast function for $l > 1$ is based on a second order difference equation with solution,

$$\tilde{y}_{T+l/T} = \alpha + \beta l$$

where α and β depend on $\tilde{y}_{T+2/T}$ and $\tilde{y}_{T+3/T}$.

If $y_T = 10$, $y_{T-1} = 9$, $\theta_1 = -1.0$, $\theta_2 = 0.2$, $\theta_3 = -0.5$, $\epsilon_T = 0.5$, $\epsilon_{T-1} = 1.0$ and $\epsilon_{T-2} = -0.2$, determine the values of $\tilde{y}_{T+l/T}$ for $l = 1, 2$ and 3 and find α and β.

3. While analysing a set of data, Mr Jones assumed that the observations had been generated by a stationary autoregressive process of order 6, namely

$$y_t = \pi_1 y_{t-1} + \cdots + \pi_6 y_{t-6} + \delta + \epsilon_t$$

(where δ is a constant), and by some appropriate procedure obtained parameter estimates

$$\hat{\pi}_1 = 0.4, \quad \hat{\pi}_2 = -0.36, \quad \hat{\pi}_3 = 0.32, \quad \hat{\pi}_4 = -0.29, \quad \hat{\pi}_5 = 0.26, \quad \hat{\pi}_6 = -0.23.$$

Show how these six parameter values could be accounted for by the values of $\pi_1, \ldots, \pi_6$ implied by the two parameters of an ARMA(1, 1) process. What would you say about the advisability of employing a six-parameter model as opposed to a two-parameter model?

Using the parameters found above for an ARMA(1, 1) model, compute $\rho(1)$, $\rho(2)$, $\rho(3)$, $\rho(4)$, and sketch the correlogram. On what basis could you have rejected an AR(1) specification immediately?

[University of York, 1980.]

4. Compute predictions for $l = 1$, 2 and 3 for the MA(2) process defined in Question 5.3. If the values given for θ_1 and θ_2 are known to be the true values, determine the MSE's of the predictions.

5. Use the von Neumann Ratio to test the hypothesis that the observations 2.0, 1.2, -0.1, -0.5, 0.3 and 0.4 are independent against the alternative that they exhibit positive serial correlation.

6. If a stationary AR(1) model with a zero mean is estimated by regression, show that $R^2 \simeq \hat{\phi}^2$.

7. Sketch the following deterministic trends and discuss how you would estimate them in models of the form (5.1):

(a) $f(t) = \exp(\alpha + \beta/t)$;
(b) $f(t) = \exp(\alpha + \beta\gamma^t)$.

8. Compute predictions and their MSE's from an AR(2) model with $\phi_1 = 0.5$ and $\phi_2 = -0.2$ for $l = 1$, 2 and 3, if $y_{T-1} = 1$, $y_T = 2$ and $\sigma^2 = 3$.

9. Write down recursive expressions for computing the derivates of ϵ_t in (7.3). Hence derive an expression for the asymptotic covariance matrix of the ML estimators, $\hat{\theta}_1$ and $\hat{\theta}_{12}$. Show that if $|\theta_1|$ is not close to unity, $\hat{\theta}_1$ and $\hat{\theta}_{12}$ are approximately uncorrelated, while the asymptotic variances of $\hat{\theta}_1$ and $\hat{\theta}_{12}$ are approximately equal to $(1 - \theta_1^2)/T$ and $(1 - \theta_{12}^2)/T$ respectively.

10. Construct the autocovariance functions for the following stationary processes:

(a) $y_t = \epsilon_t + \theta_1\epsilon_{t-1} + \theta_{12}\epsilon_{t-12}$;

(b) $y_t = (1 - \phi L^4)^{-1}(1 + \theta L^4)\epsilon_t$.

11. Derive (7.10) and (7.12).

12. Given the sample autocorrelations $r(1) = -0.34$ and $r(12) = -0.39$, estimate θ_1 and θ_{12} in (7.3).

13. Draw the autocorrelation function for the first differences of an AR(1) process with $\phi = 0.5$.

14. Derive an LM test for testing the hypothesis that $\phi_4 = 0$ in (6.5). Construct a similar test for (6.8).

15. The following model was fitted to US GNP by Nelson (1972):

$$\Delta y_t = 0.615\Delta y_{t-1} + 2.76 + \epsilon_t.$$

Determine the form of the forecast function and the eventual path of the GNP projections.

16. If $r(1) = 0.7$ and $r(2) = 0.5$ calculate the first two sample partial autocorrelations. Comment on the implications for model identification.

7
Selected Topics in Time Series Regression

1. Recursive Least Squares

Suppose that an estimator of β in a linear regression model, (1.2.6), has been calculated using the first $t - 1$ observations. The tth observation may be incorporated into the estimate without the need for inverting a cross-product matrix as implied by a direct use of the OLS formula, (1.2.13). This is carried out by means of recursive updating formulae. Let b_t be the OLS estimator based on the first t observations and let $X_t = (x_1, \ldots, x_t)'$. Then

$$b_t = b_{t-1} + (X'_{t-1}X_{t-1})^{-1}x_t(y_t - x'_t b_{t-1})/f_t \tag{1.1}$$

and

$$(X'_t X_t)^{-1} = (X'_{t-1}X_{t-1})^{-1}$$
$$- (X'_{t-1}X_{t-1})^{-1}x_t x'_t(X'_{t-1}X_{t-1})^{-1}/f_t, \tag{1.2}$$

where

$$f_t = 1 + x'_t(X'_{t-1}X_{t-1})^{-1}x_t, \qquad t = k + 1, \ldots, T. \tag{1.3}$$

The recursive updating formulae can be derived using the matrix inversion lemma given in the appendix to Chapter 4. This is done in the last sub-section where (1.1) to (1.3) emerge as special cases of the discounted recursive least squares algorithm. However, the OLS recursions can also be regarded as a special case of the Kalman filter. The classical linear regression model (1.3.6) may be identified directly with a measurement equation of the form (4.3.1) by setting $z_t = x_t$, $\xi_t = \epsilon_t$, $h_t = 1$ and $\alpha_t = \beta$. The transition equation is simply $\alpha_t = \alpha_{t-1}$ and so $T = I$ while $Q = 0$. The prediction equations are therefore trivial: (4.3.3) and (4.3.6) reduce to the identities $a_{t/t-1} = a_{t-1}$ and $P_{t/t-1} = P_{t-1}$ and the OLS updating formulae are obtained directly from (4.3.15) to (4.3.17) with $P_t = (X_t'X_t)^{-1}$.

A minimum of k observations are needed to compute an OLS estimator of β. If X_k is of full rank, the estimator based on the first k observations is

$$b_k = (X_k'X_k)^{-1}X_k'y_k^* = X_k^{-1}y_k^*, \tag{1.4}$$

where $y_k^* = (y_1, \ldots, y_k)'$. This provides a starting value and the estimators $b_{k+1}, \ldots, b_T$ can be computed with no further matrix inversions. The final estimator, b_T, is identical to the standard OLS estimator based on all T observations.

An alternative way of starting the recursions is to begin at $t = 0$ with $b_0 = 0$ and the corresponding cross-product matrix, which may be labelled $(X_0'X_0)^{-1}$, set equal to κI where κ is a large number. Although setting $b_0 = 0$ is arbitrary, setting κ equal to a large number means that the effect on estimates is negligible once k observations have been processed. If β were a random variable, these starting values could be identified with the parameters of a very weak prior distribution, but since β is fixed, starting the recursions in this way is no more than a convenient approximation.

Recursive Residuals

If the OLS estimator of β is computed recursively, a set of $T - k$ prediction errors

$$\tilde{v}_t = y_t - x_t'b_{t-1}, \qquad t = k+1, \ldots, T \tag{1.5}$$

are obtained. These have zero mean and variance $\sigma^2 f_t$. Furthermore they are uncorrelated; see EATS (Section 2.6). The standardised prediction errors

$$v_t = \tilde{v}_t/f_t^{1/2}, \qquad t = k+1, \ldots, T \tag{1.6}$$

are known as recursive residuals and they figure prominently in procedures for checking the specification and stability of regression models. A full description of these methods will be found in EATS (Section 5.2). Note that if the disturbances in (1.2.6) are normally distributed, the recursive residuals are also normal; i.e. $\epsilon_t \sim NID(0, \sigma^2)$ entails $v_t \sim NID(0, \sigma^2)$.

The recursive residuals feature in the updating formula for the residual sum of squares. This is

$$SSE_t = SSE_{t-1} + v_t^2, \qquad t = k+1, \ldots, T, \tag{1.7}$$

where

$$SSE_t = (y_t - X_tb_t)'(y_t - X_tb_t), \qquad t = k, \ldots, T. \tag{1.8}$$

Since $SSE_k = 0$, it follows that the residual sum of squares from all T observations is equal to the sum of the squares of the recursive residuals, i.e.

$$SSE_T = \sum_{t=k+1}^{T} v_t^2. \tag{1.9}$$

Discounted Recursive Least Squares

The *discounted least squares* (DLS) estimator of the parameters in a regression model is the vector $\beta = b^\dagger$, which minimises

$$S(\beta) = \sum_{t=1}^{T} \delta^{T-t}(y_t - x_t'\beta)^2, \tag{1.10}$$

where δ is a constant in the range $0 < \delta \leqslant 1$. If $\delta = 1$, $b^\dagger$ is the OLS estimator. The effect of setting δ to a value less than one is to introduce exponential weighting into the sample. Within the framework of a standard regression model, $b^\dagger$ would be the BLUE of β if it were the case that $\sigma_t^2 = \sigma^2 \delta^{t-T}$. This implies that the variances of the disturbances are greater the further back in time is the observation. However, it is not so much this assumption which underlies the use of DLS as a belief that the more remote observations are, in some sense, less reliable. The justification for a DLS estimator is therefore mainly pragmatic: if it can yield better predictions than OLS there is a strong argument for using it.

Some examples of DLS regression are given by Gilchrist (1976). He particularly stresses its application in estimating polynomial trends in models of the form (6.5.1). The use of DLS in this context corresponds to the estimation of a local, rather than a global, trend; cf. Exercise 1.2. The same technique can be used in fitting ARIMA time series models. Leskinen and Terasvirta (1976) fit a seasonal ARIMA process of order $(0, 1, 1) \times (0, 1, 1)_{12}$ to a time series on the consumption of alcoholic beverages in Finland over the period 1954 to 1974. The discount factor used was 0.9 over a twelve month period, and the estimated parameters showed some change over the coefficients obtained by a straightforward application of least squares. Interestingly enough, they found that the discounted estimates were very similar to the least squares estimates computed using the observations from 1963 to 1974 only.

The DLS estimator in a regression model can be computed recursively. The estimator of β at time t is given by

$$b_t^\dagger = \left[\sum_{j=1}^{t} \delta^{t-j} x_j x_j' \right]^{-1} \sum_{j=1}^{t} \delta^{t-j} x_j y_j, \qquad t = k, \ldots, T. \tag{1.11}$$

Noting that

$$\sum_{j=1}^{t} \delta^{t-j} x_j x_j' = P_t^{-1} = \delta P_{t-1}^{-1} + x_t x_t' \qquad (1.12)$$

and applying the matrix inversion lemma, (4A.3), gives

$$P_t = \delta^{-1} P_{t-1} - \frac{\delta^{-1} P_{t-1} x_t x_t' P_{t-1}}{\delta + x_t' P_{t-1} x_t}, \qquad t = k+1, \ldots, T. \quad (1.13)$$

This reduces to (1.2) when $\delta = 1$. The updating formula for $b_t^{\dagger}$ is obtained by first observing that

$$\sum_{j=1}^{t} \delta^{t-j} x_j y_j = \delta \sum_{j=1}^{t-1} \delta^{t-1-j} x_j y_j + x_t y_t. \qquad (1.14)$$

A little algebraic manipulation then gives

$$\begin{aligned} b_t^{\dagger} = b_{t-1}^{\dagger} &+ \frac{P_{t-1} x_t (y_t - x_t' b_{t-1}^{\dagger})}{\delta + x_t' P_{t-1} x_t} \\ &= b_{t-1}^{\dagger} + P_t x_t (y_t - x_t' b_{t-1}^{\dagger}), \qquad t = k+1, \ldots, T. \quad (1.15) \end{aligned}$$

Expression (1.1) is obtained when $\delta = 1$.

The recursive estimates can be used for tracking the parameters. The advantage of including a discount factor is that the estimates respond more quickly to a change in the structure of the model. However, this is achieved at the expense of stability. Thus there is a trade-off between sensitivity and stability, and the usual compromise is to choose a value of δ which is not too far from unity.

2. Serially Correlated Disturbances

A linear regression model with an ARMA(p, q) disturbance term may be written as

$$y_t = x_t' \beta + u_t, \qquad t = 1, \ldots, T, \qquad (2.1a)$$

$$u_t = \phi_1 u_{t-1} + \cdots + \phi_p u_{t-p} + \epsilon_t + \theta_1 \epsilon_{t-1} + \cdots + \theta_q \epsilon_{t-q}, \quad (2.1b)$$

where the notation in (2.1a) is the same as in (1.2.6). Given the specification of the model, approximate ML estimation can be carried out by a straightforward generalisation of the CSS algorithm. The technical details are described at some length in Chapter 6 of EATS. The more difficult problem of finding a suitable model for the disturbance term can be solved using the techniques developed in the previous chapter. If $\hat{\beta}$ denotes a consistent estimator of β, the residuals

$$\hat{u}_t = y_t - x_t'\hat{\beta}, \qquad t = 1, \ldots, T, \tag{2.2}$$

have the same distribution as the true disturbances in large samples and so all the asymptotic results on sample autocorrelations and diagnostic checking statistics continue to hold. The standard Box–Jenkins procedures of identification and diagnostic checking can therefore be applied directly to the residuals with the initial estimate of β given by OLS.

The above techniques can be used to handle trend plus error models of the kind defined by (6.5.1). If $f(t)$ is a polynomial, as in (6.5.2), the OLS estimator has the same asymptotic distribution as the ML estimator. It may not, however, perform as well in small samples and if the main aim of building the model is for forecasting, the ARMA disturbance term must be estimated in any case. A proof of the asymptotic efficiency of OLS for polynomial regression is given in Fuller (1976, pp. 388–93). The efficiency of the sample mean is a special case.

A model in which $f(t)$ is made up of trigonometric polynomials can also be estimated efficiently by OLS. This has obvious implications for a model in which the trend is a polynomial and the seasonal component is also deterministic; cf. (6.7.16).

Exact Maximum Likelihood Estimation*

The exact likelihood function for (2.1) can be obtained by using the methods set out in Section 5.2 and 5.3, with y_t replaced by $y_t - x_t'\beta$. This likelihood function must then be maximised with respect to β, ϕ and θ.

An alternative approach is to concentrate β out of the likelihood function by estimating it by GLS conditional on ϕ and θ. This is really a viable technique only if the inversion of the disturbance covariance matrix in (1.2.14) can be avoided. For a pure AR(p) disturbance term this is relatively straightforward; cf. EATS (Sections 6.1 and 6.4). For an MA or mixed disturbance term the solution is less obvious, but one possibility is to put the model in state space form and calculate the GLS estimator by the Kalman filter.

Suppose the model has an MA(1) disturbance term. By regarding the parameter vector β as obeying the recursion $\beta_t = \beta_{t-1}$, the state vector $\alpha_t = (\beta_t', u_t, \theta\epsilon_t)'$ can be constructed. This obeys the transition equation

$$\alpha_t = \begin{bmatrix} I & 0 \\ 0 & T* \end{bmatrix} \alpha_{t-1} + \begin{bmatrix} 0 \\ R* \end{bmatrix} \epsilon_t, \tag{2.3}$$

with

$$T* = \begin{bmatrix} 0 & 1 \\ 0 & 0 \end{bmatrix} \quad \text{and} \quad R* = \begin{bmatrix} 1 \\ \theta \end{bmatrix}; \tag{2.4}$$

cf. (4.1.7). The measurement equation is

$$y_t = z_t'\alpha_t = (x_t' \ 1 \ 0)\alpha_t. \tag{2.5}$$

The above formulation is atypical, as a state space model in that part of the state vector is non-stochastic. However, as noted in Section 4.2, this makes no difference to the Kalman filter. The only problem lies in finding suitable starting values. By analogy with the recursive least squares algorithm these could be based on the first k observations, with $a_k = (b_k' \ 0 \ 0)'$. Constructing the P_k matrix is relatively easy for MA(1) disturbances, but it can become tedious for higher order ARMA models. A more convenient approach is to begin the recursions at $t = 0$ with $a_0 = 0$ and

$$P_0 = \begin{bmatrix} \kappa I & 0 \\ 0 & P_0^* \end{bmatrix} \tag{2.6}$$

where κ is a large, but finite number and P_0^* is the matrix defined by (4.4.3). After k recursions the state vector, a_k, and the associated covariance matrix, P_k, will be almost identical to the corresponding quantities computed from the direct formulae.

As a by-product of the computation of the GLS estimator, the Kalman filter produces a set of $T - k$ standardised prediction errors

$$v_t = (y_t - z_t'a_{t/t-1})/(z_t'P_{t/t-1}z_t)^{1/2}, \qquad t = k+1, \ldots, T. \tag{2.7}$$

When ϕ and θ are known, these prediction errors, or 'generalised recursive residuals', are $NID(0, \sigma^2)$. They are identical to the recursive residuals which would have been produced if the observations had been transformed by a lower triangular matrix L satisfying $L'L = V^{-1}$. The ML estimator of σ^2, conditional on ϕ and θ, may be written in terms of the generalised recursive residuals as

$$\tilde{\sigma}^2(\phi, \theta) = T^{-1} \sum_{t=k+1}^{T} v_t^2. \tag{2.8}$$

This estimator may be used to construct the concentrated log–likelihood function

$$\log L_c(y; \phi, \theta) = -\tfrac{1}{2}T \log 2\pi - \tfrac{1}{2}T \log \tilde{\sigma}^2(\phi, \theta) - \tfrac{1}{2}\log |V|.$$

$$(2.9)$$

One problem remains in connection with (2.9) and that concerns the evaluation of $|V|$. The easiest way to approach the problem is to programme a set of 'auxiliary recursions' identical to the recursions for the matrices P_t and $P_{t/t-1}$ in the corresponding ARMA time series model. The top left-hand element in each $P_{t/t-1}$ matrix may be used to evaluate f_t for $t = 1, \ldots, T$. The determinantal term in the log–likelihood function is then given by

$$\log |V| = \sum_{t=1}^{T} \log f_t. \qquad\qquad (2.10)$$

Note that the initial matrix, P_0^*, needed to start the auxiliary recursions must be computed in any case, since it is an integral part of the GLS Kalman filter, cf. (2.6).

A two-step estimator of β can be constructed using the above algorithm. The parameters in ϕ and θ can be estimated consistently from the correlogram of OLS residuals and used to form an exact feasible GLS estimator by the Kalman filter. Two-step estimators based on approximate GLS can be derived from the CSS algorithm; see EATS (Section 6.5). Although both these estimators have the same asymptotic properties as the ML estimator, they can behave very differently in small samples. Some evidence on this matter can be found in the Monte Carlo experiments reported in Harvey and McAvinchey (1981).

A final point concerns prediction. With the GLS formulation, predictions can be made l steps ahead using the formulae given in Section 4.5. The MSE which emerges from (4.5.4) is the exact MSE conditional on ϕ and θ. In other words, an allowance for the sampling variability in the estimation of β is made automatically.

Cumulative Periodogram

When a regression model is estimated by OLS a general test against serial correlation can be based on the portmanteau statistic or its modification, (6.2.4). If attention is to be focused primarily on the first autocorrelation, $r(1)$, a bounds test based on the Durbin–Watson statistic can be carried out; see EATS (Section 6.3).

The use of the cumulative periodogram as a general test against serial correlation was described in Section 6.2. When the periodogram ordinates in (6.2.10) are constructed from OLS residuals, Durbin (1969) has shown that a bounds test is possible. This is relatively easy to carry out. In place of a single line, $s = c_0 + i/n$, two lines are drawn. When the path of s_i crosses the upper line, the null hypothesis is rejected against the alternative of an excess of low frequency. If it fails to cross the lower line, the null hypothesis is not rejected, while crossing the lower line but not the upper line is taken to be inconclusive. Carried out in this way the test corresponds to a Durbin–Watson bounds test against positive serial correlation. A two-sided bounds test may also be carried out by constructing a corresponding pair of lines below the 45° line.

If $n^* = \frac{1}{2}(T - k)$, the two outer lines are

$$s = \pm c_0 + i/n^*,$$

while the inner lines are

$$s = \pm c_0 + [i - \tfrac{1}{2}(k - 1)]/n^*,$$

for $\frac{1}{2}k - \frac{1}{2} \leqslant i \leqslant n^*$. Thus the only tables needed are those for carrying out the standard cumulative periodogram test. No additional tables of upper and lower bounds are needed since the difference between the inner and outer lines depends only on $k - 1$, the number of degrees of freedom attributable to the regression. The bounds are exact when T and k are both odd, but if this is not the case Durbin (1969, p. 10) notes that '... the amount of approximation is slight and should be negligible in practice unless n^* is small'.

Example 1 Figure 7.1 shows the cumulative periodogram for the

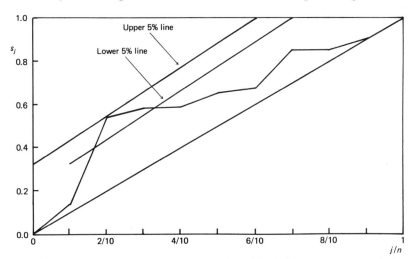

Figure 7.1 *Cumulated Periodogram s_j of Residuals from Regression of Consumption on Wages and Profits for USA 1921–41 (21 observations)*

Source: Durbin (1969, p. 13).

OLS residuals from the regression of consumption on profits and wages as originally carried out by Klein (1950, pp. 74–5, 135). In this case $T = 21$ and $k = 3$ and so $n^* = 9$. For a one-sided test at the 5% level of significance $c_0 = 0.32538$, and this figure is used in constructing the lines on the diagram. On inspecting figure 7.1 it will be observed that s_i rises initially, indicating that low frequencies are dominant in the residuals. However, although s_i crosses the lower line, it fails to cross the upper line, and the test is inconclusive.

3. Time-Varying Parameters

Suppose that the structure of a regression equation changes from one period to the next. A viable formulation of such a model can be obtained by assuming that the vector of parameters, β_t, is generated by a stochastic process. There are basically three classes of model which can be adopted. In the first, the parameters are assumed to vary randomly about a fixed, but unknown, mean. This is known as a *random coefficient* model. It is relatively easy to handle, and a number of applications have appeared in the literature; see, for example, Theil (1971, pp. 622–27). However, if parameters are to be regarded as stochastic, it seems reasonable to suppose that they will change gradually over time rather than in a haphazard fashion. This suggests a more general formulation in which β_t is generated by a multivariate ARMA process. The assumption that this process is stationary ensures that β_t moves around a fixed mean and hence the term *'return to normality'* is sometimes used to characterise the model.

The third approach to modelling time-varying parameters assumes that they are generated by a stochastic process which is non-stationary. An important example of this kind of behaviour arises when the parameters follow a multivariate *random walk*. Because the parameters are no longer constrained to have a fixed mean, the model can gradually evolve over time. The values taken at the end of the sample period can be very different from those at the beginning, and a model exhibiting this kind of property may be very attractive if it is felt that radical changes have taken place in the underlying relationship.

Maximum likelihood estimation procedures are developed below for all three classes of time-varying parameters. For the return to normality and random walk models, the Kalman filter provides the most convenient route for deriving the likelihood function. Two-step estimators can also be constructed within this framework.

Random Coefficients

In the random coefficient model the parameter vector at time t is drawn from a multivariate normal distribution with mean $\bar{\beta}$ and covariance matrix Q. Thus

$$y_t = x_t'\beta_t, \tag{3.1a}$$

$$\beta_t = \bar{\beta} + \epsilon_t, \qquad \epsilon_t \sim NID(0, Q), \tag{3.1b}$$

for all t. If the first element in x_t is always unity, it is unnecessary to add a disturbance term to (3.1a) since such a term is implicitly contained in (3.1b), attached to the coefficient of the constant term.

Substituting (3.1b) into (3.1a) yields

$$y_t = x_t'\bar{\beta} + u_t, \qquad t = 1, \ldots, T, \tag{3.2}$$

where $u_t = x_t'\epsilon_t$. Model (3.2) is a standard heteroscedastic regression model, in which the disturbances are normally and independently distributed with mean zero and variance,

$$\sigma_t^2 = x_t'Qx_t, \qquad t = 1, \ldots, T. \tag{3.3}$$

Matters are simplified considerably if Q is diagonal. Expression (3.3) then reduces to

$$\sigma_t^2 = q_1 + q_2 x_{2t}^2 + \cdots + q_k x_{kt}^2, \qquad t = 1, \ldots, T, \tag{3.4}$$

where $q_1, \ldots, q_k$ are the diagonal elements of Q.

Formulating the model as (3.2) and (3.4) suggests a relatively straightforward two-step estimation procedure. Estimators of $q_1, \ldots, q_k$ are first obtained by regressing $\hat{u}_t^2$, the square of the tth OLS residual, on $x_{2t}^2, \ldots, x_{kt}^2$. These estimators, $\hat{q}_1, \ldots, \hat{q}_k$, are then used to construct a weighted least squares (WLS) estimator of $\bar{\beta}$,

$$\tilde{b} = \left[\sum_{t=1}^{T} \hat{\sigma}_t^{-2} x_t x_t' \right]^{-1} \sum_{t=1}^{T} \hat{\sigma}_t^{-2} x_t y_t, \tag{3.5a}$$

where $\sigma_t^2 = q_1 + q_2 x_{2t}^2 + \cdots + q_k x_{kt}^2$. The WLS estimator, $\tilde{b}$, is a feasible GLS estimator of $\bar{\beta}$. Under suitable regularity conditions $\hat{q}_1, \ldots, \hat{q}_k$ are consistent and $\tilde{b}$ is asymptotically efficient.

Maximum likelihood estimates of $\bar{\beta}$ and $q = (q_1, \ldots, q_k)'$ can be computed using the method of scoring. The information matrix is block diagonal with respect to $\bar{\beta}$ and q, and it is not difficult to show that the scoring iteration for $\bar{\beta}$ reduces to a WLS regression of the form (3.5a). As regards q, the scoring equation for q also reduces to a WLS regression as

$$q^* = \hat{q} + \left[E\left(-\frac{\partial^2 \log L}{\partial q \, \partial q'} \right) \right]^{-1} \frac{\partial \log L}{\partial q}$$

$$= \left[\sum_{t=1}^{T} \hat{\sigma}_t^{-4} w_t w_t' \right]^{-1} \sum_{t=1}^{T} \hat{\sigma}_t^{-4} w_t (y_t - x_t'\hat{\bar{\beta}})^2, \tag{3.5b}$$

where $w_t = (1 \ \ x_{2t}^2, \ldots, x_{kt}^2)'$.

Return to Normality Model*

Consider equation (3.1a) and suppose that the parameters are generated by a stationary multivariate AR(1) process,

$$\beta_t - \bar{\beta} = \Phi(\beta_{t-1} - \bar{\beta}) + \epsilon_t, \tag{3.6}$$

where $\epsilon_t \sim NID(0, Q)$ and Φ is a $k \times k$ matrix of parameters. Since Φ contains k^2 unknown parameters it will generally be taken to be diagonal, in which case the stationarity condition is simply that no element should have an absolute value greater than, or equal to, one. Allowing the parameters to follow a scheme of this kind enables them to change slowly over time, around a fixed mean $\bar{\beta}$. Schaefer *et al.* (1976) investigated this type of parameter variation in connection with a share's market risk and found a good deal of support for it. An earlier application will be found in Rosenberg (1973).

The random coefficient model (3.1b) is obtained as a special case of (3.6) by setting $\Phi = 0$. However, the introduction of an autoregressive element leads to what is potentially a much more useful representation. Furthermore, any stationary multivariate ARMA process for β_t can be handled within this framework. A multivariate ARMA process can be expressed as a multivariate AR(1) process, and the likelihood function evaluated by the Kalman filter. This technique can also be applied in the present context. However, although β_t can, in principle, be modelled by any ARMA process, the practical problems of identifying and estimating structures more complicated than (3.6) are likely to be considerable.

Substituting (3.6) into (3.1a) leads to a regression model of the form (3.2) in which

$$u_t = x_t'(\beta_t - \bar{\beta}) = x_t'\delta_t, \qquad t = 1, \ldots, T. \tag{3.7}$$

Although u_t has zero expectation, it differs from the disturbance term in the random coefficient model in that it is serially correlated as well as heteroscedastic. The elements of the covariance matrix of $(u_1, \ldots, u_T)'$, are given by

$$E(u_t u_s) = x_t' \Gamma(\tau) x_s, \qquad t, s = 1, \ldots, T \qquad (3.8)$$

where $\tau = t - s$ and $\Gamma(\tau)$ is the autocovariance matrix of $(\beta_t - \bar{\beta})$ at lag τ. With both Φ and Q diagonal, (3.8) becomes

$$E(u_t u_s) = \sum_{i=1}^{k} x_{it} x_{is} q_i \phi_i^{\tau} / (1 - \phi_i^2), \qquad t, s = 1, \ldots, T. \qquad (3.9)$$

Expression (3.9) forms the basis for a *two-step* estimation procedure. Replacing $E(u_t u_s)$ by $\hat{u}_t \hat{u}_s$ for $\tau = 0$ and $\tau = 1$ suggests two regressions, one of $\hat{u}_t^2$ on $x_{2t}^2, \ldots, x_{kt}^2$ and the other of $\hat{u}_t \hat{u}_{t-1}$ on $x_{2t} x_{2,t-1}, \ldots, x_{kt} x_{k,t-1}$. This yields $2k$ estimated coefficients from which it is possible to solve for the $2k$ unknown parameters in Φ and Q. Thus if $\hat{\gamma}_{i,0}$ denotes the ith regression coefficient in the equation for $\tau = 0$, and $\hat{\gamma}_{i,1}$ denotes the corresponding coefficient for $\tau = 1$, ϕ_i may be estimated by $\hat{\phi}_i = \hat{\gamma}_{i,1} / \hat{\gamma}_{i,0}$. Once estimates of the ϕ_i's have been obtained, estimating the q_i's is straightforward. Estimating $q_1, \ldots, q_k$ and $\phi_1, \ldots, \phi_k$ in this way represents a natural generalisation of the first stage in the two-step estimator in the random coefficient model. The second step, as in the random coefficient case, is to use these estimates to construct a feasible GLS estimator of $\bar{\beta}$. If the model is cast in state space form, the GLS estimator can be computed using the Kalman filter without the need to invert, or even explicitly construct, the covariance matrix defined by (3.8).

The appropriate state space formulation is similar to that given for the regression model with ARMA disturbances in Section 2. The state vector is $\alpha_t = (\bar{\beta}_t' \delta_t')'$, where $\bar{\beta}_t = \bar{\beta}$ for all t while $z_t = (x_t' x_t')'$. The transition equation is as in (2.3), but with $T^* = \Phi$ and $R^* = I$. It will usually be convenient to start the recursions at $t = 0$ with P_0 of the form (2.6). The matrix P_0^* is equal to $\Gamma(0)$, but when Φ and Q are diagonal, $\Gamma(0)$ is also diagonal with ith diagonal element equal to $q_i / (1 - \phi_i^2)$.

There are two approaches to maximum likelihood estimation. The first is based on the GLS estimator and the idea is to 'concentrate out' $\bar{\beta}$ from the likelihood function. The covariance matrix of ϵ_t is re-defined as $\sigma^2 \bar{Q}$ where the top left-hand element of $\bar{Q}$ is unity. Thus the GLS Kalman filter recursions are independent of σ^2 and this parameter can also be concentrated out of the likelihood function leaving a function of the form (2.9) to be maximised with respect to Φ and $\bar{Q}$.

In the second approach no attempt is made to concentrate the likelihood function. Since $y_t^{\dagger} = y_t - x_t' \bar{\beta} = x_t' \delta_t$ is a stationary series, the likelihood function can be decomposed in terms of T innovations as in (1.2.21). The appropriate state space formulation is

$$y_t^{\dagger} = x_t' \delta_t,$$
$$\delta_t = \Phi \delta_{t-1} + \epsilon_t, \qquad t = 1, \ldots, T. \tag{3.10}$$

The initial conditions for the state vector, δ_t, are automatically available since it is a stationary process. Thus $\delta_0 \sim N(0, \Gamma(0))$ where $\Gamma(0)$ has the simple diagonal form noted at the end of the previous paragraph but one.

Random-Walk Parameters

In the random-walk parameter model,

$$y_t = x_t' \beta_t + \epsilon_t, \qquad t = 1, \ldots, T, \tag{3.11a}$$

the vector β_t is generated by the process

$$\beta_t = \beta_{t-1} + \eta_t, \tag{3.11b}$$

where $\eta_t \sim NID(0, \sigma^2 Q)$. This model differs from the autoregressive scheme of the previous sub-section in that the parameters are allowed to evolve over time. Because the process is non-stationary, β_t has no fixed mean and so the model is able to accommodate fairly fundamental changes in structure.

The $k \times k$ matrix Q determines the extent to which the parameters are allowed to vary. If $Q = 0$, the model collapses to a classical regression model since if $\beta_t = \beta_{t-1}$, for all t, the parameters are constant. If Q is p.d., all k parameters will be time-varying. There is no reason why this should necessarily be the case, although in the discussion below it will be convenient to assume that if one parameter is stochastic so are the rest.

As it stands, a likelihood function cannot be defined for the random-walk model. However, following Cooley and Prescott (1976), a possible approach is to regard β_T as being fixed. By repeated substitution each β_t may be expressed in terms of β_T. Thus, for example,

$$\beta_T = \beta_{T-1} + \eta_T = \beta_{T-2} + \eta_{T-1} + \eta_T$$

and so

$$y_{T-2} = x_{T-2}' \beta_T + [\epsilon_{T-2} - x_{T-2}'(\eta_{T-1} + \eta_T)].$$

In general,

$$y_t = x_t' \beta_T + \zeta_t, \qquad t = 1, \ldots, T$$

where

$$\zeta_t = \epsilon_t - x_t'(\eta_{t+1} + \eta_{t+2} + \cdots + \eta_T). \tag{3.12}$$

Written in this way the model may be viewed within the standard GLS framework. The parameter vector, β_T, is fixed, while the disturbances have zero mean and covariance matrix, $\sigma^2 V$. The ijth element of V is given by

$$v_{ij} = [\min(T-i, T-j)] \cdot x_i'Qx_j + \delta_{ij}, \tag{3.13}$$

where δ_{ij} is zero for $i \neq j$ and unity for $i = j$.

The GLS estimator of β_T may therefore be computed for a given value of Q. Thus β_T and σ^2 may be concentrated out of the likelihood function and ML estimation can be carried out by maximising an expression of the form (2.9) with respect to the elements of Q. Cooley and Prescott (1976) show how this may be done without repeated inversions of the $T \times T$ matrix V.

Choosing β_T to be fixed is essentially arbitrary. In fact, Cooley and Prescott take β_{T+1} to be fixed. However, rather than pursuing the relative merits of conditioning on different β_t's, we next consider an entirely different approach, based on the Kalman filter. This is a much more natural technique to adopt in these circumstances. The model may be put into the state space format of Section 4.3 simply by changing the notation: $\alpha_t = \beta_t$, $z_t = x_t$, $\xi_t = \epsilon_t$ and so on. The only problem lies in forming starting values.

An initialisation problem arises with the OLS recursions, but is solved by using the first k observations to construct starting values. The same approach may be adopted here. The first k observations are expressed in the form

$$y_t = x_t'\beta_k + \zeta_t, \qquad t = 1, \ldots, k,$$

where ζ_t is defined as in (3.12) with $T = k$. The MMSE of β_k is then

$$b_k = (X_k'V_k^{-1}X_k)^{-1}X_k'V_k^{-1}y_k^* = X_k^{-1}y_k^*, \tag{3.14}$$

where X_k and y_k^* are as defined in (1.4) and $\sigma^2 V_k$ is the covariance matrix of $(\zeta_1, \ldots, \zeta_k)'$. The covariance matrix of $b_k - \beta_k$ is given by $\sigma^2 P_k$ where

$$P_k = (X_k'V_k^{-1}X_k)^{-1}. \tag{3.15}$$

Taken together, b_k and P_k form the starting values for the Kalman filter. Conditional on Q, the recursions yield the MMSE of β_T once all the observations have been processed. If $Q = 0$, $P_k = (X_k'X_k)^{-1}$ and the whole scheme reduces to the recursive least squares algorithm. The explicit calculation of b_k and P_k can be avoided by starting the recursions at $t = 0$ with $a_0 = 0$ and $P_0 = \kappa I$, where κ is a large scalar number.

There is no need to take β_k as fixed in deriving (3.14) and (3.15). In fact, in order to obtain a likelihood function from the Kalman filter, it becomes necessary to fix b_k. This means that β_k has a proper prior distribution, i.e. $\beta_k \sim N(b_k, \sigma^2 P_k)$, and so it follows from the prediction error decomposition that the likelihood function for $y_{k+1}, \ldots, y_T$ may be expressed in terms of the Kalman filter innovations, i.e.

$$\log L(y_{k+1}, \ldots, y_T) = -\tfrac{1}{2}(T - k)\log 2\pi - \tfrac{1}{2}(T - k)\log \sigma^2$$

$$-\tfrac{1}{2} \sum_{t=k+1}^{T} \log f_t - \tfrac{1}{2}\sigma^{-2} \sum_{t=k+1}^{T} v_t^2/f_t.$$

(3.16)

At first sight it might appear that (3.16) is more limited than the likelihood function obtained from the GLS approach, since the latter is defined for all T observations. However, this is not the case, since although $y_1, \ldots, y_k$ are fixed, all parameter vectors are taken to be random.

The scalar parameter, σ^2, may be concentrated out of the likelihood function. The ML estimator of Q is then obtained by minimising the function

$$\log L_c = \log \tilde{\sigma}^2(Q) + (T - k)^{-1} \sum_{t=k+1}^{T} \log f_t$$

where

$$\tilde{\sigma}^2(Q) = (T - k)^{-1} \sum_{t=k+1}^{T} v_t^2/f_t.$$

Example 1   Using annual observations for the period 1916–1966, Garbade (1977) estimated the following demand for money equation by OLS:

$$\Delta \widehat{\log M1}_t = 0.0020 - 0.0833\,\Delta \log Rcp_t + 0.4947\Delta \log y_t$$

$$R^2 = 0.360 \quad s = 0.051 \quad DW = 1.43 \tag{3.17}$$

The variables in (3.17) are the narrow money supply per capita ($M1$), the commercial paper yield (Rcp), which is taken to represent the rate of interest, and real income per capita (y). Garbade then proceeded to estimate the model under the assumption that the parameters had been generated by a random walk process, (3.11b). The matrix Q was taken to be diagonal and ML estimates of the three elements, q_{11}, q_{22} and q_{33}, were computed by exploiting the Kalman filter in the way described above.[1] This gave the following results: $\tilde{q}_{11} = 0.1982$, $\tilde{q}_{22} = 2.6791$, $\tilde{q}_{33} = 5.4574$ and $\tilde{\sigma} = 0.0337$.

Having estimated Q, Garbade computed *smoothed* estimates of the β_t's using the algorithm described in Section 4.5. The graphs on p. 62 of his paper show substantial and well defined fluctuations in β_{1t} and β_{2t}, but the income elasticity, β_{3t}, appears to stabilise in the mid 1930s at around 0.5. However, it may be that these parameter variations are symptomatic of a misspecified dynamic model rather than an indication of a fundamental change in the process generating the data. Indeed, it has been argued that tracking the parameters in this way is an extremely useful diagnostic checking procedure; see Salmon (1980).

Discounting Past Observations

In Section 1, discounted least squares was proposed as a method of estimating regression models when there is doubt about the stability of the parameters over time. An alternative approach is to estimate the parameters by the Kalman filter algorithm associated with the random-walk parameter model. The matrix Q is fixed *a priori* according to the discount factor which it is desired to bring into the estimation of each parameter.

If x_t consists only of a constant term, (3.11) reduces to (6.5.32). The optimal forecasts for the latter model take the form of an exponentially weighted moving average. Thus a discounting scheme based on (3.11) represents a natural generalisation of a standard time series technique.

4. Regression in the Frequency Domain

The classical linear regression model may be set up in the frequency domain by applying a finite Fourier transform to the dependent and independent variables. This creates T sets of observations which are indexed not by time, but by frequency. *Spectral regression* is then carried out by regressing the transformed dependent variable on the transformed independent variables. There are a number of reasons for doing this. One is to permit the application of the technique

---

[1]There is a minor difference between Garbade's method and that described in the text, in that in Garbade's algorithm P_k is not calculated using (3.15). The method set out here was originally proposed in Harvey and Phillips (1976b).

known as *band spectrum regression*, in which regression is carried out in the frequency domain with certain wavelengths omitted. This has a number of interesting applications with regard to dealing with seasonality and errors in the explanatory variables. For example, if the observations show a strong seasonal pattern, it may be advantageous to try to discount any seasonal effects in the relationship by omitting the transformed observations which lie on and around the seasonal frequencies. With respect to the errors in variables problem, it has been suggested that the adverse effects on the least squares estimator might be mitigated by deleting the high frequency observations from a frequency domain regression. Since economic variables tend to be slowly-changing, whereas errors of measurement are typically taken to be 'white noise', the distortion in the variables will be relatively more severe at higher frequencies. The lower frequencies, on the other hand, are relatively robust to this type of measurement error, and so restricting spectral regression to these frequencies may prove to be advantageous. Further discussion of these ideas will be found in Engle (1974a; 1978b).

A second major reason for being interested in spectral regression is that if the disturbances are serially correlated in the time domain, they will be approximately uncorrelated in the frequency domain. However, the elimination of serial correlation is obtained at the expense of the introduction of heteroscedasticity. Nevertheless, this has some advantages in that the variances of the transformed disturbances are closely related to the spectral density of the original disturbances. This has implications both for estimation and testing. In particular, it suggests a spectral estimator which is *non-parametric* in the sense that it does not rely on a particular specification for the process generating the disturbance term.

Transformation to the Frequency Domain

The generalised regression model may be written in matrix terms as

$$y = X\beta + u \tag{4.1}$$

where u is a $T \times 1$ vector of disturbances with mean zero and covariance matrix, $\sigma^2 V$; cf. (1.2.12). Let W be the transpose of the $T \times T$ Fourier matrix, Z, defined in (3.2.19). Pre-multiplying the observations in (4.1) by W gives

$$\dot{y} = \dot{X}\beta + \dot{u}, \tag{4.2}$$

where $\dot{y} = Wy$, $\dot{X} = WX$ and $\dot{u} = Wu$. Although the Fourier matrix in (3.2.19) is defined for the case of T being even, the transformation matrix for T odd is of a similar form and the appropriate formulae when T is odd may easily be deduced by referring back to Section 3.2.

Transforming the observations in this way leads to a frequency domain interpretation of the linear regression model. However, (4.2) contains no more, and no less, information than does (4.1). In fact, if the disturbance term in (4.1) is white noise, the disturbance term in (4.2) will also be white noise. This follows immediately from the orthogonality of W since

$$E(\dot{u}\dot{u}') = E(Wuu'W') = \sigma^2 WW' = \sigma^2 I.$$

Applying OLS to the transformed observations in the frequency domain therefore gives the BLUE of β, but this estimator is identical to the OLS estimator computed from the untransformed observations. A formal demonstration is straightforward:

$$b = (\dot{X}'\dot{X})^{-1}\dot{X}'\dot{y} = (X'W'WX)^{-1}X'W'Wy = (X'X)^{-1}X'y. \quad (4.3)$$

The interesting point about the transformation to the frequency domain is that successive pairs of observations correspond to different frequencies. Certain frequency components may therefore be omitted simply by dropping the appropriate transformed observations from the data set. This is known as band spectral regression. Since the transformed observations are real, any standard regression package may be used to carry out the calculations, and all the results appropriate to the classical linear regression model may be applied without modification. Thus, for example, when $u_t \sim NID(0, \sigma^2)$, F-statistics may be used to test whether the observations corresponding to certain frequency components obey the same model as the remaining observations. Such tests are identical to analysis of covariance tests in a conventional regression model.

Further insight into what is happening in spectral regression may be obtained by considering the effect of the Fourier transform on each of the variables. For simplicity, it is convenient to assume that the model contains only one regressor, so that in the frequency domain,

$$\dot{y}_s = \beta \dot{x}_s + \dot{u}_s, \qquad s = 1, \ldots, T. \tag{4.4}$$

The periodogram of the observations on the original explanatory variable is then given by

$$p_j = \dot{x}_{2j}^2 + \dot{x}_{2j+1}^2, \qquad \text{for } j = 1, \ldots, n-1, \tag{4.5}$$
$$p_n = \dot{x}_T^2.$$

Apart from a factor of proportionality, the periodogram is the same as the sample spectral density, and from the point of view of further development it is easier to work with the second of these two quantities. Thus $I_x(\lambda_j) = p_j/4\pi$ for $j = 1, \ldots, n - 1$ and $I_x(\lambda_n) = p_n/2\pi$, while $I_x(\lambda_0)$ may be set equal to $\dot{x}_1^2/2\pi = T^{1/2}\bar{x}/2\pi$. The real part of the sample cross spectrum, $I_{xy}^*(\lambda_j)$ is defined in a similar fashion, with $I_{xy}^*(\lambda_j) = (\dot{x}_{2j}\dot{y}_{2j} + \dot{x}_{2j+1}\dot{y}_{2j+1})/4\pi$ for $j = 1, \ldots, n - 1$. The OLS estimator of β may therefore be written as

$$b = \sum_{s=1}^{T} \dot{x}_s \dot{y}_s \bigg/ \sum_{s=1}^{T} \dot{x}_s^2$$

$$= \sum_{j=0}^{m} \psi_j I_{xy}^*(\lambda_j) \bigg/ \sum_{j=0}^{m} \psi_j I_x(\lambda_j), \tag{4.6}$$

where $\psi_j = 0.5$ for $j = 0$ and $j = n$, and $\psi_j = 1$ elsewhere. However, because the imaginary part of the sample cross-spectrum is an odd function, it disappears under a two-sided summation. Thus (4.6) can be re-expressed as

$$b = \sum_{j=-n}^{n-1} I_{xy}(\lambda_j) \bigg/ \sum_{j=-n}^{n-1} I_x(\lambda_j), \tag{4.7}$$

where $I_{xy}(\lambda_j)$ is the sample cross-spectrum. With an odd number of observations, the summation in (4.7) is from $-n$ to n.

In the general case when the model contains $k > 1$ independent variables, $I_x(\lambda_j)$ is replaced by $I_{xx}(\lambda_j)$, a $k \times k$ matrix of sample spectra and cross-spectra. Similarly $I_{xy}(\lambda_j)$ is a $k \times 1$ vector of cross-spectral elements and so (4.7) becomes

$$b = \left[\sum_{j=-n}^{n-1} I_{xx}(\lambda_j) \right]^{-1} \sum_{j=-n}^{n-1} I_{xy}(\lambda_j). \tag{4.8}$$

Serial Correlation

Suppose that the disturbances in a regression model have zero mean, but are generated by a stationary stochastic process. The effect of the transformation into the frequency domain is to produce a vector of disturbances, $\dot{u}$, with covariance matrix

$$\dot{\Gamma} = E(\dot{u}\dot{u}') = WE(uu')W' = W\Gamma W'.$$

Because the original disturbances are stationary, Γ is a Toeplitz matrix, and it can be shown that a matrix of this form is approximately diagonalised by a finite Fourier transform. If $\dot{\gamma}_{ij}$ denotes the ijth element of $\dot{\Gamma}$, then

$$\dot{\gamma}_{ij} \simeq \begin{cases} 2\pi f_u(\lambda_j), & i = j \\ 0, & i \neq j \end{cases}; \tag{4.9}$$

where $f_u(\lambda)$ is the spectral density of the disturbance term. This expression is exact if the process generating the disturbances is regarded as being circular. Without this assumption, the approximation can be shown to be negligible if T is large; cf. Fishman (1969).

If (4.9) were exact and the spectral density of the disturbances were known, the GLS estimator of β could be computed in the frequency domain by weighted least squares. In the more usual case when $f_u(\lambda)$ is unknown, it must first be estimated. Provided this is done consistently, it is possible to apply standard asymptotic theory to show that the resulting feasible GLS estimator will be asymptotically efficient.

The additional computational burden imposed by the need to estimate $f_u(\lambda)$ is not very great. Given a consistent estimator of β, say the OLS estimator, the corresponding residuals will be consistent estimators of the true disturbances. If the true disturbances were known, $f_u(\lambda)$ could be estimated consistently by smoothing their periodogram. A similar smoothing procedure applied to the periodogram calculated from consistent estimates of the disturbances will likewise produce a consistent estimate of $f_u(\lambda)$. The computations required to form an estimator of $f_u(\lambda)$ in this way are relatively insignificant because if $e = y - Xb$ is the vector of OLS residuals in the time domain, it follows that $\dot{e} = \dot{y} - \dot{X}b = We$. The periodogram of the OLS residuals may therefore be obtained *directly* from the frequency domain residuals. Thus, having transformed the observations into the frequency domain, an asymptotically efficient estimator of β may be computed in two steps. At the first step a consistent estimator of β is obtained by applying OLS to the transformed model, (4.2). The power spectrum of the original disturbances is estimated directly from the resulting frequency domain residuals, simply by squaring them, adding together successive pairs, and smoothing. An efficient estimator is then obtained using weighted least squares. If desired, this procedure may be iterated, with a view to improving the small sample properties of the estimator. The costs involved are likely to be small relative to those incurred in carrying out the original Fourier transforms.

The GLS estimator may be written in much the same way as the OLS estimator in (4.7). For a single regressor equation, the feasible GLS estimator is

$$\tilde{b} = \sum_{j=-n}^{n-1} \hat{f}_u^{-1}(\lambda_j) I_{xy}(\lambda_j) \Big/ \sum_{j=-n}^{n-1} \hat{f}_u^{-1}(\lambda_j) I_x(\lambda_j), \tag{4.10}$$

where $\hat{f}_u(\lambda_j)$ is a consistent estimator of $f_u(\lambda)$. However, since $\hat{f}_u(\lambda)$ is obtained by smoothing, it is also legitimate to consider smoothing $I_x(\lambda_j)$ and $I_{xy}(\lambda_j)$. Furthermore, a continuity argument suggests that a reduction in the number of points at which the spectra and cross-spectra are estimated will have little effect on the estimator. If N is the lag length employed in the spectral estimator and $\lambda_i = 2\pi i/N$, an alternative to (4.10) is

$$\tilde{b}^\dagger = \sum_{i=-N}^{N-1} \hat{f}_u^{-1}(\lambda_i)\hat{f}_{xy}(\lambda_i) \Big/ \sum_{i=-N}^{N-1} \hat{f}_u^{-1}(\lambda_i)\hat{f}_x(\lambda_i). \tag{4.11}$$

Under suitable conditions, $\tilde{b}^\dagger$ has the same asymptotic distribution as $\tilde{b}$. A choice between the two will largely rest on computational convenience, although the result for white noise disturbances has led Engle and Gardner (1976) to argue strongly in favour of (4.10).

The interpretation of the feasible GLS estimator in (4.11) suggests that

$$\text{Avar}(\tilde{b}^\dagger) = T^{-1}\left(\frac{1}{2\pi} \int_{-\pi}^{\pi} f_u^{-1}(\lambda)f_x(\lambda)d\lambda\right)^{-1}. \tag{4.12}$$

An expression of this kind is usually easier to evaluate than a corresponding formula in the time domain. Thus there are often theoretical as well as practical advantages to working in the frequency domain. This point will not be pursued here, but the interested reader may consult Engle and Gardner (1976) or Nicholls and Pagan (1977) for further details.

5. Non-Parametric Treatment of Serial Correlation

If the process generating the disturbance term in a regression model is misspecified, there may be a considerable loss in efficiency. The most common example of a misspecification of this kind is to assume that u_t is generated by an AR(1) process, when a more general ARMA model would be appropriate. In these circumstances the resulting estimator may actually be more inefficient than OLS. In other words, a feeble attempt at modelling the disturbance term may actually be worse than making no correction at all. Some evidence on this matter may be found in Engle (1974b) and Engle and Gardner (1976).

One way of avoiding these problems is to use a spectral estimator such as (4.10) or (4.11). Although an estimator computed in this way is asymptotically efficient, it is *non-parametric* in the sense that the process generating the disturbance term need not be specified. An alternative way of constructing a non-parametric estimator of β is to model the disturbance by an $AR(P)$ process, where P is chosen in such a way that P/T remains constant as $T \to \infty$. This is essentially a time domain analogue of (4.10). The estimator is relatively easy to compute. The OLS residuals, e_t, are regressed on their lagged values, $e_{t-1}, \ldots, e_{t-P}$, to produce consistent (and efficient) estimators of the AR parameters, $\phi_1, \ldots, \phi_P$. The estimator of β is then obtained by regressing

$$y_t - \hat{\phi}_1 y_{t-1} - \cdots - \hat{\phi}_P y_{t-P}, \qquad t = P + 1, \ldots, T$$

on

$$x_t - \hat{\phi}_1 x_{t-1} - \cdots - \hat{\phi}_P x_{t-P}, \qquad t = P + 1, \ldots, T.$$

This 'autoregressive least squares' estimator is sometimes referred to as $ALS(P)$.

If the spectrum of u_t is estimated by the autoregressive procedure described at the end of Section 3.6, the relationship between the time and frequency domain estimators will be very close. The correspondence will not be exact because of the approximation implied by (4.9), and the smoothing needed to obtain $\hat{f}_u(\lambda)$. Although this may result in significant differences in certain special cases, the two estimators will, for most practical purposes, be the same.

Non-parametric estimators are asymptotically efficient, but this does not necessarily mean that they are effective in small samples. The Monte Carlo results presented in Engle and Gardner (1976) suggest that for $T = 100$, the variance of the spectral estimator will, at best, be twice that of the asymptotic variance. This leaves plenty of scope for a more efficient small sample estimator based on a suitably parsimonious representation of the disturbance term. In fact, even if the disturbance term is misspecified, it is still quite possible for the resulting estimator to be more efficient than a non-parametric estimator in small samples. Engle and Gardner again provide evidence on this phenomenon, with one particular set of experiments showing ALS(1) dramatically outperforming the spectral estimator for $T = 100$, even though the disturbance was actually generated by an $AR(2)$ process. Supporting evidence comes from Gallant and Goebel (1976) where, for a particular (non-linear) regression model, the ALS(2) estimator had a smaller MSE than the spectral estimator for an underlying MA(4) disturbance term.

The above results suggest that non-parametric regression cannot always be recommended, particularly if it is carried out in the frequency domain. One possible compromise may be to adopt an ALS approach with P chosen by a simple goodness of fit rule such as the AIC. This can be implemented in a semi-automatic fashion, thereby avoiding the need to identify a suitable ARMA process.

Notes

Section 2   Harvey and Phillips (1979).
Section 3   See the papers in *The Annals of Economic and Social Measurement* (1973), Harvey (1978b) and Pagan (1980). A Monte Carlo study on the properties of ML and two-step estimators in the return to normality model is reported in Harvey and Phillips (1981).
Section 4   Hannan (1963), Harvey (1978a).
Section 5   Akaike (1969), Amemiya (1973).

Exercises

1. Explain how you would obtain approximate ML estimates in a regression model with MA(1) disturbances by using the CSS algorithm. Write down recursive expressions for computing the first derivatives.
   Derive the large sample covariance matrix of the estimators.
2. Explain how you would make predictions of future values of the dependent variable in a regression model with ARMA(1, 1) disturbances, given the appropriate values of the explanatory variables. Write down an expression for computing the MSE of the predictions conditional on the estimated parameters in the model.
3. Suppose that the parameters ϕ and θ in a regression model with ARMA(1, 1) disturbances are known. Show how the GLS estimator of β can be computed by setting up the model in state space form. Explain how the state space model can be used to make predictions for future values of y_t, given the corresponding values of x_t. Set up 95% prediction intervals. Are these intervals exact, given that β had to be estimated?
4. A regression model with a slowly changing seasonal component may be written in the form

$$y_t = s_t + x_t'\beta + \epsilon_t$$
$$s_t = s_{t-s} + \eta_t, \qquad t = 1, \ldots, T,$$

where ϵ_t and η_t are independent, normally distributed white noise processes. Explain how you would estimate a model of this form, and how you would make predictions of future values of y_t, given knowledge of the relevant x_t's. What would be the consequences for estimation and prediction if s_t were deterministic rather than stochastic?
5. Suppose that observations are generated by the model

$$y_t = \alpha + \beta x_t + \epsilon_t, \qquad t = 1, \ldots, T,$$

where α is a non-zero parameter. If the model is incorrectly specified without the constant term, show that the OLS estimator of β will be biased. If β were assumed to be time-varying in the manner suggested by (3.11b), and the model was estimated on this basis (again without the constant term) do you think the predictions would be any better? Does your answer have any implications for misspecified models in general?

Answers to Selected Exercises

Chapter 1. — 1. $\hat{\beta}_{71} = 2.56$, $\hat{\alpha}_{71} = 26.3 + 2.56(5) = 39.14$, $\hat{\beta}_{72} = 2.58$, $\hat{\alpha}_{72} = 41.86$ etc.

Chapter 2. — 1. (i) Yes; (ii) No; (iii) No, No; (iv) Yes; roots are $\frac{1}{2} \pm \frac{1}{2}i$.
    2. 0.267, -0.167.
    4. (b) This is the autocorrelation function in figure 2.4.
    6. Yes; there is a common factor of $(1 + 0.8L)$. The model reduces to an ARMA(1, 1) process.
    7. $\rho_{yx}(\tau) = 0$ for $\tau < 0$ and $\tau > 2$.
    9. $|\theta(L)| = (1 + \theta L)(1 - L)$.

Chapter 3. — 1. Yes; see Anderson (1971, pp. 403–5).
    4. $\text{Ph}(\lambda) = \tan^{-1}[-q(\lambda)/c(\lambda)]$.
    8. Moving average filter cuts down high frequencies, but has no phase shift. Difference filter has a phase shift and removes trend, $A(\lambda) = 1 - \exp(-i7\lambda)$.
    10. $\rho(1) = -0.5$, $\rho(2) = 0$, $\rho(3) = -0.5\phi$, $\rho(\tau) = \phi\rho(\tau - 4)$ for $\tau \geqslant 4$; $f(\lambda) = 2(1 - \cos \lambda)/(1 + \phi^2 - 2\phi \cos 4\lambda)$.
    11. $\hat{y}_t = \dfrac{19}{20} + \dfrac{1}{2} \cos \dfrac{2\pi}{4} t + \dfrac{5}{2} \sin \dfrac{2\pi}{4} t - \dfrac{1}{4}(-1)^t$.

    Standardised coefficients of cosine and sine terms (distributed as $N(0, 1)$) are 0.94 and 4.71 respectively. Standardised coefficient of last term is -0.67.

Chapter 4. — 1. $\bar{P} = 4.20$.
    2. The prediction equation, (3.19a), becomes: $a_{t/t-1} = T_t a_{t-1} + c_t$. Equation (3.19b) is unchanged, as are the updating equations.

Chapter 5. — 3. $S(\theta_1, \theta_2) = 54.322$ and so $\tilde\sigma^2 = 54.322/5$. MSE's for $l = 1$, 2 and 3 are σ^2, $1.25\sigma^2$ and $1.29\sigma^2$ respectively.

4. From the Yule–Walker equations, $\tilde\phi_1 = 1.11$, $\tilde\phi_2 = -0.39$ and $\tilde\sigma^2 = 0.306$. Peak in spectrum at $\lambda = \cos^{-1}(1.11/2\sqrt{0.39}) = 0.48$.

7. $\tilde\phi = 0.6$, $\tilde\mu = 7.2$, $\mathrm{Avar}(\tilde\mu) = T^{-1}c(0)[1 + r(1)]/[1 - r(1)]$; see Box and Jenkins (1976, p. 195).

8. cf. (7.11).

9. LM test — regress ϵ_t on derivatives with respect to ϕ, θ_0 and θ_1.

Chapter 6. — 1. If $\rho(\tau) = 0$ for $\tau \geqslant 2$, then from (3.2), $\mathrm{avar}[r(2)] = 0.022$. Reject $Ho : \rho(2) = 0$ at 5% level of significance.

2. 10.8, 11.2, 11.35, $\alpha = 10.9$, $\beta = 0.15$.

7. (a) Regress $\log y_t$ on $1/t$; (b) non-linear least squares, e.g. Gauss–Newton.

9. Box and Jenkins (1976, p. 319).

12. $\hat\theta_1 = -0.39$, $\hat\theta_{12} = -0.48$.

13. $\rho(1) = -(1 - \phi)/2 = -0.25$; $\rho(\tau) = \rho(1)\phi^{\tau-1}$.

14. $\mathrm{LM} = T \cdot r^2(4)$. Tested as χ_1^2.

16. $\tilde\phi(1) = r(1) = 0.7$. From (5.2.5), $\tilde\phi(2) = 0.02$. Suggests AR(1) specification.

Chapter 7. — 4. Apply seasonal difference operator to y_t and x_t and model the disturbance term by a seasonal MA(s) process of the form $u_t = \xi_t + \theta\xi_{t-s}$, where ξ_t is white noise; cf. Section 6.7. If s_t is deterministic, $\theta = -1$. Use full ML and finite sample prediction; cf. starred sub-section in Section 6.7 (p. 182).

References

Akaike, H. (1969), Fitting autoregressive models for prediction, *Annals of the Institute of Statistical Mathematics*, 21, pp. 243–247.

Akaike, H. (1974), A new look at the statistical model identification, *IEEE Transactions on Automatic Control*, AC-19, pp. 716–723.

Amemiya, T. and Wu, R. Y. (1972), The effect of aggregation on prediction in the autoregressive model, *Journal of the American Statistical Association*, 67, pp. 628–632.

Amemiya, T. (1973), Generalized least squares with an estimated auto-covariance matrix, *Econometrica*, 41, pp. 723–732.

Anderson, B. D. O. and Moore, J. B. (1979), *Optimal Filtering*, Prentice-Hall.

Anderson, T. W. (1971), *The Statistical Analysis of Time Series*, John Wiley, New York.

Ansley, C. F. (1979), An algorithm for the exact likelihood of a mixed autoregressive moving average process, *Biometrika*, 66, pp. 59–65.

Ansley, C. F. and Newbold, P. (1980), Finite sample properties of estimators for autoregressive moving average models, *Journal of Econometrics*, 13, pp. 159–184.

Baillie, R. T. (1979), Asymptotic prediction mean squared error for vector autoregressive models, *Biometrika*, 66, pp. 675–678.

Beach, C. M. and MacKinnon, J. G. (1978), A maximum likelihood procedure for regression with auto-correlated errors, *Econometrica*, 46, pp. 51–58.

Box, G. E. P. and Jenkins, G. M. (1976), *Time Series Analysis: Forecasting and Control*, revised edition, Holden-Day, San Francisco.

Box, G. E. P. and Pierce, D. A. (1970), Distribution of residual autocorrelations in autoregressive integrated moving average time series models, *Journal of the American Statistical Association*, 65, pp. 1509–1526.

Brewer, K. R. W. (1973), Some consequences of temporal aggregation and systematic sampling for ARMA and ARMAX models, *Journal of Econometrics*, 1, pp. 133–154.

Burman, J. P. (1980), Seasonal adjustment by signal extraction, *Journal of the Royal Statistical Society, Series A*, 143, pp. 321–337.

Chan, W. Y. T. and Wallis, K. F. (1978), Multiple time series modelling: another look at the mink–muskrat interaction, *Applied Statistics*, 27, pp. 168–175.

Cleveland, W. S. and Tiao, G. C. (1976), Decomposition of seasonal time series: a model for the census X-11 program, *Journal of the American Statistical Association*, 71, pp. 581—587.

Cooley, T. F. and Prescott, E. C. (1976), Estimation in the presence of stochastic parameter variation, *Econometrica*, 44, 167—184.

Crowder, M. H. (1976), Maximum likelihood estimation for dependent observations, *Journal of the Royal Statistical Society, Series B*, 38, pp. 45—53.

Davidson, J. (1981), Problems with the estimation of moving average processes, *Journal of Econometrics* (forthcoming).

Dent, W. T. and Min, A. S. (1978), A Monte Carlo study of autoregressive integrated moving average processes, *Journal of Econometrics*, 7, pp. 23—55.

Duncan, D. B. and Horn, S. D. (1972), Linear dynamic regression from the viewpoint of regression analysis, *Journal of the American Statistical Association*, 67, pp. 815—821.

Durbin, J. (1969), Tests for serial correlation in regression analysis based on the periodogram of least-squares residuals, *Biometrika*, 56, pp. 1—15.

Engle, R. F. (1974a), Band spectrum regression, *International Economic Review*, 15, pp. 1—11.

Engle, R. F. (1974b), Specification of the disturbance term for efficient estimation, *Econometrica*, 42, pp. 135—146.

Engle, R. F. (1978a), Estimating structural models of seasonality (with discussion), in A. Zellner (ed.), *Seasonal Analysis of Economic Time Series*, Bureau of the Census, Washington DC, pp. 281—308.

Engle, R. F. (1978b), Testing price equations for stability across spectral frequency bands, *Econometrica*, 46, pp. 869—881.

Engle, R. F. and Gardner, R. (1976), Some finite sample properties of spectral estimators of a linear regression, *Econometrica*, 44, pp. 149—166.

Fishman, G. S. (1969), *Spectral Methods in Econometrics*, Harvard University Press.

Fuller, W. A. (1976), *Introduction to Statistical Time Series*, John Wiley, New York.

Fuller, W. A. and Hasza, D. P. (1978), Properties of prediction for autoregressive time series, Dept. of Statistics, Iowa State University (unpublished paper).

Galbraith, R. F. and Galbraith, J. F. (1974), On the inverse of some patterned matrices arising in the theory of stationary time series, *Journal of Applied Probability*, 11, pp. 63—71.

Gallant, A. R. and Goebel, J. J. (1976), Nonlinear regression with autocorrelated errors, *Journal of the American Statistical Association*, 7, pp. 961—967.

Garbade, K. (1977), Two methods for examining the stability of regression coefficients, *Journal of the American Statistical Association*, 72, pp. 54—63.

Gardner, G., Harvey, A. C. and Phillips, G. D. A. (1980), An algorithm for exact maximum likelihood estimation by means of Kalman filtering, *Applied Statistics*, 29, 311—322.

Gilchrist, W. (1976), *Statistical Forecasting*, John Wiley, New York.

Godfrey, L. G. (1978), Testing against general autoregressive and moving average error models when the regressors include lagged dependent variables, *Econometrica*, 6, pp. 1293—1302.

Godolphin, E. J. (1976), On the Cramér—Wold factorization, *Biometrika*, 63, pp. 367—380.

Goldberg, S. (1958), *Difference Equations*, John Wiley, New York.

Granger, C. W. J. and Hatanaka, M. (1964), *Spectral Analysis of Economic Time Series*, Princeton University Press.

Granger, C. W. J. and Morris, M. J. (1976), Time series modelling and interpretation, *Journal of the Royal Statistical Society, Series A*, 139, pp. 246—257.

Granger, C. W. J. and Newbold, P. (1977), *Forecasting Economic Time Series*, Academic Press, New York.

Griffiths, L. J. and Prieto-Diaz, R. (1977), Spectral analysis of natural seismic events using autoregressive techniques, *IEEE Transactions on Geo-Science Electronics*, GE-15, pp. 13—25.

Hannan, E. J. (1963), Regression for time series, in M. Rosenblatt (ed.), *Time Series Analysis*, John Wiley, New York, pp. 14—37.

Hannan, E. J. (1969), The identification of vector mixed autoregressive moving average systems, *Biometrika*, 56, pp. 223—225.

Hannan, E. J. (1970), *Multiple Time Series*, John Wiley, New York.

Hannan, E. J. (1980), The estimation of the order of an ARMA process, *Annals of Statistics*, 8, pp. 1071—1081.

Harrison, P. J. (1967), Exponential smoothing and short-term sales forecasting, *Management Science*, 13, pp. 821—842.

Harrison, P. J. and Stevens, C. F. (1976), Bayesian forecasting (with discussion), *Journal of the Royal Statistical Society, Series B*, 38, pp. 205—247.

Hart, B. I. (1942), Significance levels for the ratio of the mean square successive difference to the variance, *Annals of Mathematical Statistics*, 13, pp. 445—447.

Harvey, A. C. (1978a), Linear regression in the frequency domain, *International Economic Review*, 19, pp. 507—512.

Harvey, A. C. (1978b), The estimation of time-varying parameters from panel data, *Annales de l'INSEE*, 30—31, pp. 203—226.

Harvey, A. C. (1980), On comparing regression models in levels and first differences, *International Economic Review*, 21, pp. 707—720.

Harvey, A. C. (1981a), *The Econometric Analysis of Time Series*, Philip Allan.

Harvey, A. C. (1981b), *Finite Sample Prediction and Overdifferencing*, Discussion Paper, London School of Economics.

Harvey, A. C. and McAvinchey, I. D. (1981), On the relative efficiency of various estimators of regression models with moving average disturbances, in G. Charatsis (ed.), *Proceedings of European Meeting of the Econometric Society*, Athens, 1979.

Harvey, A. C. and Pereira, P. L. V. (1980), The estimation of dynamic models with missing observations, Paper presented to *World Congress of the Econometric Society*, Aix-en-Provence, August 1980.

Harvey, A. C. and Phillips, G. D. A. (1976a), The maximum likelihood estimation of autoregressive-moving average models by Kalman filtering, University of Kent (unpublished paper).

Harvey, A. C. and Phillips, G. D. A. (1976b), Testing for stochastic parameters in regression, Paper presented at the *European Meeting of the Econometric Society*, Helsinki, August 1976.

Harvey, A. C. and Phillips, G. D. A. (1977), A comparison of estimators in the ARMA(1, 1) model, University of Kent (unpublished paper).

Harvey, A. C. and Phillips, G. D. A. (1979), The estimation of regression models with autoregressive-moving average disturbances, *Biometrika*, 66, pp. 49—58.

Harvey, A. C. and Phillips, G. D. A. (1981), The estimation of regression models with time-varying parameters, in *Proceedings of a Symposium in Honour of Oskar Morgenstern*, Vienna, May 1980.

Hasza, D. P. (1980), The asymptotic distribution of the sample autocorrelations for an integrated ARMA process, *Journal of the American Statistical Association*, 75, pp. 349–352.

Hosking, J. R. M. (1980), The multivariate portmanteau statistic, *Journal of the American Statistical Association*, 75, pp. 602–608.

Jazwinski, A. H. (1970), *Stochastic Processes and Filtering Theory*, Academic Press, New York.

Jones, R. H. (1980), Maximum likelihood fitting of ARMA models to time series with missing observations, *Technometrics*, 22, pp. 389–395.

Kalman, R. E. (1960), A new approach to linear filtering and prediction problems, *Transactions ASME Journal of Basic Engineering*, 82, pp. 35–45.

Kang, K. M. (1975), A comparison of estimators for moving average processes, Australian Bureau of Statistics (unpublished paper).

Kendall, M. G. (1973), *Time Series*, Griffin.

Klein, L. R. (1950), *Economic Fluctuations in the United States, 1921–41*, John Wiley, New York.

Kuznets, S. S. (1961), *Capital and the American Economy: Its Formation and Financing*, National Bureau of Economic Research, New York.

Leskinen, E. and Terasvirta, T. (1976), Forecasting the consumption of alcoholic beverages in Finland, *European Economic Review*, 8, pp. 349–369.

Ljung, G. M. and Box, G. E. P. (1978), On a measure of lack of fit in time series models, *Biometrika*, 66, pp. 67–72.

Nelson, C. R. (1972), The prediction performance of the FRB–MIT–PENN model of the U.S. Economy, *American Economic Review*, 62, pp. 902–917.

Nelson, C. R. (1974), The first order moving average process, *Journal of Econometrics*, 2, pp. 121–141.

Nelson, C. R. (1976), Gains in efficiency from joint estimation of systems of autoregressive-moving average processes, *Journal of Econometrics*, 4, pp. 331–348.

Nelson, C. R. and Shea, G. S. (1979), Hypothesis testing based on goodness-of-fit in the moving average time series model, *Journal of Econometrics*, 10, pp. 221–226.

Nerlove, M., Grether, D. M. and Carvalho, J. L. (1979), *Analysis of Economic Time Series*, Academic Press, New York.

Nicholls, D. F. and Pagan, A. R. (1977), Specification of the disturbance for efficient estimation — an extended analysis, *Econometrica*, 45, pp. 211–217.

Osborn, R. R. (1976), Maximum likelihood estimation of moving average processes, *Annals of Economic and Social Measurement*, 5, pp. 75–87.

Pagan, A. (1980), Some identification and estimation results for regression models with stochastically varying parameters, *Journal of Econometrics*, 13, pp. 341–363.

Parzen, E. (1969), Multiple time series modelling, in P. R. Krishaiah (ed.), *Multivariate Analysis, Vol. II*, Academic Press, New York, pp. 389–409.

Poskitt, D. S. and Tremayne, A. R. (1980), Testing the specification of a fitted autoregressive-moving average model, *Biometrika*, 67, pp. 359–363.

Poskitt, D. S. and Tremayne, A. R. (1981), *A Time Series Application of the Use of Monte Carlo Methods to Compare Statistical Tests*, Discussion Paper, University of York.

Rosenberg, B. (1973), Random coefficient models: the analysis of a cross section of time series by stochastically convergent parameter regression, *Annals of Economic and Social Measurement*, 2, pp. 399–428.

Roy, R. (1977), On the asymptotic behaviour of the sample autocovariance function for an integrated moving average process, *Biometrika*, 64, pp. 419–421.

Salmon, M. (1980), Structural constancy in model selection, Paper presented to *World Congress of the Econometric Society*, Aix-en-Provence, August 1980.

Sargan, J. D. and Bhargava, A. (1980), *Maximum Likelihood Estimation of Regression Models with First Order Moving Average Errors when the Root Lies on the Unit Circle*, Discussion paper, London School of Economics.

Schaefer, S. *et al.* (1975), Alternative models of systematic risk, in E. Elton and M. Gruber (eds), *International Capital Markets: An Inter and Intra Country Analysis*, North-Holland Publishing Co., pp. 150—161.

Schweppe, F. (1965), Evaluation of likelihood functions for Gaussian signals, *IEEE Transactions on Information Theory*, 11, pp. 61—70.

Shenton, L. R. and Johnson, W. L. (1965), Moments of a serial correlation coefficient, *Journal of the Royal Statistical Society, Series B*, 27, pp. 308—320.

Silvey, S. D. (1970), *Statistical Inference*, Chapman and Hall.

Theil, H. (1971), *Principles of Econometrics*, John Wiley, New York.

Theil, H. and Goldberger, A. S. (1961), On pure and mixed statistical estimation in economics, *International Economic Review*, 2, pp. 65—78.

Wallis, K. F. (1974), Seasonal adjustment and relations between variables, *Journal of the American Statistical Association*, 69, pp. 18—31.

Wilson, G. T. (1973), The estimation of parameters in multivariate time series models, *Journal of the Royal Statistical Society, Series B*, 35, pp. 76—85.

Yamamoto, T. (1976), A note on the asymptotic mean square error of predicting more than one step ahead using the regression method, *Applied Statistics*, 25, pp. 123—127.

Author Index

Subject Index